D1597382

Developments in Economic Geology, 5

PRINCIPLES OF INDUCED POLARIZATION FOR GEOPHYSICAL EXPLORATION

Further titles in this series

1. *I.L. ELLIOTT and W.K. FLETCHER*
GEOCHEMICAL EXPLORATION 1974

2. *P.M.D. BRADSHAW*
CONCEPTUAL MODELS IN EXPLORATION GEOCHEMISTRY
The Canadian Cordillera and Canadian Shield

3. *G.J.S. GOVETT and M.H. GOVETT*
WORLD MINERAL SUPPLIES
Assessment and Perspective

4. *R.T. SHUEY*
SEMICONDUCTING ORE MINERALS

Developments in Economic Geology, 5

PRINCIPLES OF INDUCED POLARIZATION FOR GEOPHYSICAL EXPLORATION

J.S. SUMNER

Laboratory of Geophysics, Departments of Geosciences and Mining and Geological Engineering, The University of Arizona, Tucson, Arizona, U.S.A.

ELSEVIER SCIENTIFIC PUBLISHING COMPANY
Amsterdam — Oxford — New York 1976

ELSEVIER SCIENTIFIC PUBLISHING COMPANY
335 Jan van Galenstraat
P.O. Box 211, Amsterdam, The Netherlands

Distributors for the U.S.A. and Canada:

ELSEVIER NORTH-HOLLAND INC.
52, Vanderbilt Avenue
New York, N.Y. 10017

ISBN: 0-444-41481-9

With 165 illustrations and 8 tables

Printed in The Netherlands

To my wife, Nancy

PREFACE

A direct means of indicating the presence of concealed mineralization would be the ultimate in exploration. In many ways the induced-polarization (IP) technique comes close to achieving this goal. It has been remarkably successful in locating concealed bodies containing even small percentages of metallic-luster minerals.

Induced polarization as a subject has not been presented to any great extent in college courses, and many exploration men have therefore not had the benefit of training in this technique. Because of the rapid growth in this field, there is a serious lag in the published literature, even though use of the method is becoming rather widespread. Not much is available on IP in an intermediate-level treatment and some of the basic research data are still in confidential company files. Hopefully this book will help fill the information gap that is growing in this expanding field.

Because of its costs, IP surveying can be a major expense item in a mineral exploration budget. It therefore behooves persons in exploration to have more than a casual acquaintanceship with aspects of the technique.

Unlike most other geophysical exploration methods, induced polarization is usually not useful for anything but prospecting for metallic-luster minerals. Other uses may develop in good time, for example, in petroleum exploration well logging, but at the present the goal of most IP surveys is to discover or to delineate an economic mineral deposit.

Although more than 90% of the geophysically interpreted, drill-tested IP anomalies seem to be explained by the presence of metallic-luster minerals, such as sulfides, the sulfide minerals themselves may not be even indirectly related to ore mineralization. Much subore grade pyrite has been found with IP surveys. Other exploration methods and especially geological reasoning must be used to sort out the noneconomic IP sources.

This volume is written for geologists, engineers, and anyone who is interested in technical aspects of mineral exploration. It is expected that the reader's background will include some familiarity with undergraduate college-level physics, mathematics, and chemistry.

The object of this book is to review IP phenomena, measuring equipment, and techniques, and to describe the interpretation of results under a variety of geologic environments and operating conditions. There often is a choice that can be made concerning types and means of measurements and interpretation of field results. Sometimes the results of these decisions are critical to the success of the exploration program. Material in this book is intended to assist the exploration man to become more familiar with induced polarization in order that his work becomes more efficient and effective.

JOHN S. SUMNER

Tucson, Arizona

ACKNOWLEDGMENTS

The writing of this book would not have been possible without the generous assistance of many people. Students in the exploration geophysics courses at The University of Arizona have made contributions and have offered helpful criticisms, most of which have been incorporated and have improved the quality of the manuscript. In particular, W.A. Sauck, K.L. Zonge, and J.C. Wynn have allowed me to use the results of their research into the induced-polarization phenomena. Tucson geophysicists C.L. Elliot, C. Ludwig, D. McLean, W.E. Heinrichs, Jr., and G. Weiduwilt have provided background material and data on equipment. Mining company geophysicists A. Brant, B. Cromak, S.I. Gaytan, A.T. Hauck III, M. Nabighian, P.M. Wright, J.D. Corbett, and R. Myers have supported research projects at The University of Arizona, the results of which are incorporated in this book. Geophysical consultants P.G. Hallof, N. Patterson, and H.O. Seigel have given encouragement and have provided me with their data. Professors S.H. Ward of the University of Utah and T.R. Madden of the Massachusetts Institute of Technology have given of their time in reviewing portions of the manuscript, and Dr. N.F. Ness of NASA has made his EM coupling studies available to me. Helen R. Hauck has assisted me in converting classroom lecture notes into a manuscript, creating order from chaos.

Induced polarization still has its controversial aspects even among the best qualified geophysicists, but I have tried to present the different concepts without bias. Nevertheless, the views expressed are indeed my own, and I accept the responsibility for errors of emphasis, omission, and fact that may eventually become apparent.

JOHN S. SUMNER

Tucson, Arizona, 1975

TABLE OF CONTENTS

Chapter 1

INTRODUCTION

Induced polarization (IP) is a current-stimulated electrical phenomenon observed as delayed voltage response in earth materials. It has practical importance as a method of exploring the subsurface for buried mineral deposits. In geophysical measurements, induced polarization refers to a resistive blocking action or electrical polarization in earth materials, the process being most pronounced in fluid-filled pores next to metallic minerals. The IP effect is therefore observed to be strongest near rocks containing metallic-luster minerals. However, the exact relationship between IP response and amount of mineralization is complex.

The primary advantage of the IP method is its capability under favorable conditions to detect the presence of even very small amounts of metallic-luster minerals. According to company reports, disseminated sulfide bodies with as little as 0.5% by volume 'metallics' have been successfully identified as being the cause of an IP anomaly.

Measurable effects due to differences in the physical properties of rocks can be found using gravitation, magnetic, and seismic geophysical exploration methods. Knowing the characteristics of the measuring system, data can be interpreted over the range of variations of a physical property. Anomalous differences in these apparent values, however, are often small compared with the normal variation of the fundamental property, therefore non-IP methods suffer both from the difficulties in constructing sensitive and accurate measuring instruments and from problems in interpreting the subsurface situations from field results.

In the IP method, the presence of an indication in itself is sufficient to define an anomaly, rather than a difference in measurement level. That is to say, the IP phenomenon usually has a low background or threshold value above which measurements can be simply called 'anomalous'. Thus there is an advantage to the IP method in that it yields a more definitive interpretation. This distinction is in marked contrast to the closely related resistivity and self-potential prospecting methods, which are seldom as direct.

The magnitude of IP response generally increases with the amount of mineralization up to the point where massive sulfide bodies can give large effects. However, this economically interesting type of geologic occurrence is usually detectable by other less expensive electrical geophysical methods, such as electromagnetic, resistivity, or self-potential methods. Hence, induced polarization is a particularly useful technique in areas of disseminated mineralization where other geophysical exploration methods are much less effective.

A HISTORICAL SUMMARY OF IP RESEARCH

Conrad Schlumberger was the first to describe conclusively an IP response (*polarisation provoquée*) in his classic monograph *Étude sur la prospection électrique du sous-sol*, published in 1920, and he properly deserves credit for recognizing the phenomenon. The initial discovery is disclosed in his German patent (No. 269,928) of 1912. He implied that uninteresting background effects tended to obliterate response due to mineralization, so his accounts of the observation tended to discourage further investigation for many years.

The name 'induced polarization' has its origin in translation of the Schlumberger papers. Some geophysicists consider this term not completely appropriate in English because 'induction' is a name that is used for electrical processes in which electrical fields are propagated through space without conducting materials being present. It has been commonly accepted that 'induced polarization' refers to the induced electrical polarization of more or less conductive rocks. Ward (1967) has suggested that 'induced electrical polarization' would be a better descriptive name for the phenomenon because a magnetic polarization can be induced in rocks. However, the term 'induced electrical polarization' still would not exclude dielectric effects or any electrostatic behavior of dry rocks. Nevertheless, induced polarization is now accepted both in name and in fact.

Despite some rather confusing descriptions of polarization-like behavior of earth materials which can be found in the published literature in the three decades that followed the early Schlumberger report, there were no further significant published developments on the field exploration method until rediscovery by Brant and his group at Newmont Exploration Limited in 1946 sparked an application report by Bleil (1953) and a comprehensive series of articles under the guidance of MIT's Madden (1957—1959).

Research in the Soviet Union on induced polarization in petroleum well logging started in 1941, according to Dakhnov's (1959) descriptions of the method. Later development work in Russia has been carried out by Komarov (1969).

The early developmental work at Newmont is well summarized in a series of fundamental articles edited by Wait (1959), entitled *Overvoltage Research and Geophysical Applications.* These papers will be often referred to in this book. Brant's brief historical review gives appropriate testimony to the significant pioneering work and the rapid progress in research stimulated by exploration successes by his group.

Madden and his co-workers at MIT contributed four comprehensive reports to the U.S. Atomic Energy Commission between 1957 and 1959. These articles treat a variety of basic theoretical and laboratory research topics with emphasis on IP effects due to nonmetallic causes. Further work in induced polarization has also been accomplished by Madden's former students.

For a while it seemed that Vacquier and his associates (1957) at New Mexico Institute of Mining and Technology had been able to utilize IP techniques for finding groundwater resources because of related polarization effects in dirty sands and water-saturated clays. These workers developed a theory, laboratory techniques, and field instrumentation. They proposed that membrane phenomena were a dominant cause of induced polarization. Their sensitive pulse equipment was probably also influenced by electromagnetic coupling effects between ground circuits due to very low resistivities in their test areas.

In the years between 1955 and 1960, several universities, mining companies, and government agencies became interested in IP research because of the success of the method in exploring for metallic minerals. Students in engineering geophysics at the University of California at Berkeley have accomplished much since 1958 in IP research, especially in investigating phenomena involving rock properties. Michigan Technological University, Missouri School of Mines, the University of Utah, and the U.S. Geological Survey were engaged in IP research over these years. Base-metal mining companies, including Kennecott Copper, Anaconda, American Smelting and Refining Co., and Phelps Dodge, started their own research groups to increase their in-house geophysical exploration capability. Geophysical contractors, such as McPhar Geophysics, Huntec, Geoscience, Heinrichs Geoex, and Seigel Associates, became established in IP technology and built their own equipment, often at considerable developmental expense. In more recent years, the Schlumberger group has renewed activity in this field, and good research is being accomplished by Japanese and Russian geophysicists.

IP research and surveying since 1960 have maintained their earlier growth rate, as indicated on Table I. Induced polarization is the leading geophysical

TABLE I

Growth trend in IP activity

Year	Research		Ground surveys	
	man-months	expenditures	man-months	expenditures
1961	99.5	$108,970	484.8	$ 702,673
1962	98.0	116,500	446.3	728,100
1963	128.0	141,100	796.4	1,382,100
1964	225.6	303,100	705.8	1,876,100
1965	219.6	359,000	1,082.1	2,460,700
1966	282.0	409,200	1,565.2	5,109,900
1967	288.0	463,650	1,511.9	5,681,780
1968	296.4	625,200	2,404.5	6,735,315
1969	Reporting procedure changed			7,200,000

From annual activity summaries published yearly in *Geophysics* by the Geophysics Activity Committee of the Society of Exploration Geophysicists.

ground prospecting method, present expenditures accounting for about a third of the total. Expenditures, both in dollars and in manpower involved in research and in field surveying, have continued to increase at the rate of about 25% per year. Of course, this growth rate will not continue indefinitely, but as long as mankind's mineral needs keep growing there will be a demand for improvements in exploration ideas.

THE IP EFFECT

The principal cause of induced polarization in mineralized rocks is a current-induced electron transfer reaction between electrolyte ions and metallic-luster minerals. When metallic minerals block or are next to electrolyte-filled pore paths and an electric current flows through the rock, an electrochemical over-potential (overvoltage) builds up at the interface between the electron-conducting mineral and the pore solution. These electrochemical forces that oppose current flow are described as polarizing the interface, and the increase in voltage required to drive current through the interface is called the overvoltage.

Most of the electric current that is passed through unmineralized rock is carried by the water in fractures and pore spaces because adjacent rock-forming minerals are very poor conductors of electricity. Even though highly conductive metallic-luster sulfide minerals may lie in the fractures, their presence only serves to disrupt the current flow because of the attendant polarization phenomenon. Not all of the induced blocking potential is immediately present at the start of current flow, so initially the rock resistivity is lower; then polarization resistance builds up to the point where the overvoltage reaction reaches a maximum. The electrical behavior is similar to the process of charging a leaky capacitor. The final amount of polarization is proportional to the amount of current passing through the rock.

In order for induced polarization to take place it is necessary that there be fluid-filled pores in a material. An electrolyte must be present that is in contact with the metallic mineral. With no pores and no electrolyte, there would be no appreciable IP effect and the resistivity of the material would be quite high. From the observed fact that natural mineralized rocks in place display an IP effect, we know that rocks in place are at least slightly fractured and porous and that they are wet.

The IP phenomenon is measured by passing a controlled inducing current through a substance and observing resultant voltage changes with time or with variations of inducing frequency. When the inducing current is turned off, the primary voltage almost immediately drops to a secondary response level and then the transient decay voltage diminishes with time (Fig.1.1). Observation of this secondary voltage or decay phenomenon is one means of measuring the polarization of a material.

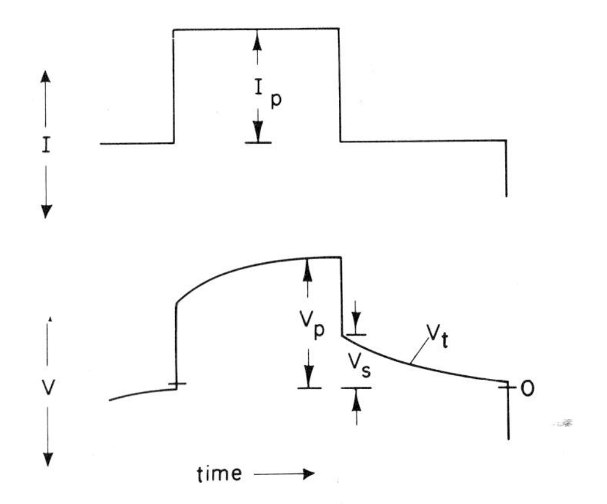

Fig.1.1. The IP pulse or time-method waveforms, ideally due to a long-period pulse, showing the induced primary current I_p being detected as a maximum primary voltage V_p. When current is turned off, voltage drops to a secondary level V_s and the transient voltage V_t decays with time. A theoretical measure of chargeability, M, is $M = V_s/V_p$.

Since it takes time to build up polarization, the voltage and resistance across the polarized zone decrease with increasing frequency so that meaningful voltage measurements can also be made at different frequencies (Fig.1.2). The difference in voltage response at different frequencies is therefore also related to the polarization properties of materials.

IP decay-voltage amplitudes are ordinarily related directly to the earlier applied electric current. This proportionality would indicate that conditions of response in the natural system are approximately in equilibrium with conditions of stimulation, at least over the range of practical field measurements. The greater time lag of the delayed IP decay-voltage response in comparison with the short decay times of truly inductive electromagnetic effects implies that there are long conductive electrical paths or perhaps there is an intermediate phase involving the exchange of electrical energy.

The amount of polarization is proportional to and linear with the amount of current that is passing through a material. Also, the polarization voltages seen on cycled charge and discharge curves each bear the same relationship with time to the extent that IP charging and decay waveforms are symmetrically identical in form. This system symmetry, also described as linearity, is analogous to the mechanical input and output properties of action and reaction of a quasi-elastic system in which energy loss is proportional to energy input. In electrical terms, system linearity means that the input (current) and output (voltage) parameters are proportional. In other words, the impedance of a linear circuit or system is independent of the applied current or voltage. The impedance of a linear system is characteristic of the system and may be complex and is sometimes called a transfer function. This is illustrated in Fig.1.3.

The changing voltage of the IP decay curve, which will be referred to as IP

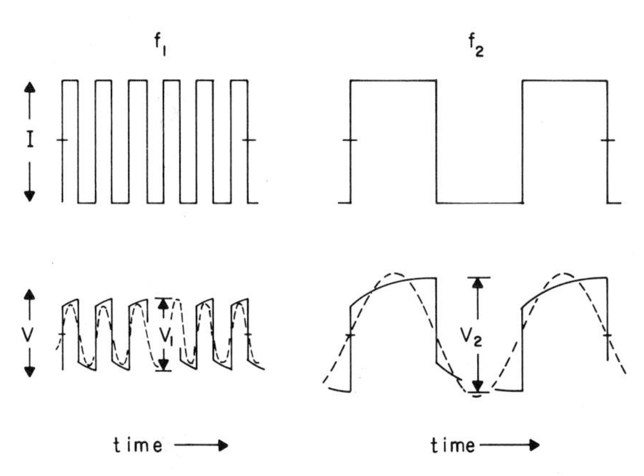

Fig.1.2. IP frequency-method waveforms, showing a controlled, constant inducing current I at frequencies f_1 and f_2 being detected as voltages V_1 and V_2, where $V_1 < V_2$. The common measure of frequency effect is $FE = (V_2 - V_1)/V_1$ or simply $FE = \Delta V_f/V_1$. The dashed line is the sinusoidal filtered voltage.

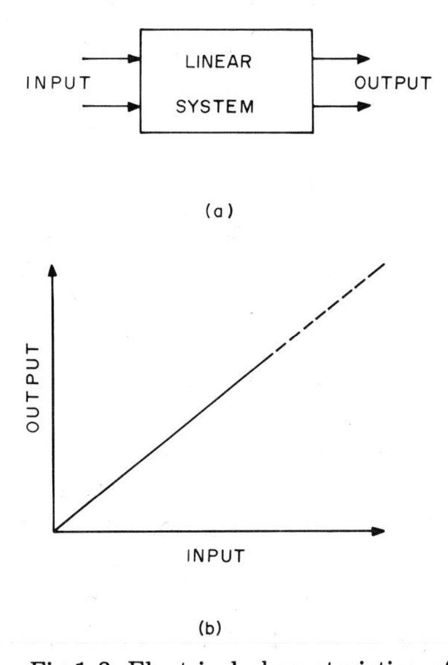

Fig.1.3. Electrical characteristics of a linear system. The ratio of output voltage to input current, by Ohm's law, is the transfer impedance of the system. (a) Block diagram of a linear system or circuit showing input and output, which may be complex waveforms. (b) The plot of input against output of a linear circuit. Earth materials behave in a linear fashion at low current densities.

response, is due to an internal current within and around the polarized material from ions returning to their equilibrium positions acting under the influence of previously induced electric fields.

The IP effect is similar to the dielectric property of materials, and indeed there is a class of dielectrics into which can be placed electrically polarizable earth materials. Conventional insulating dielectric substances owe their polarization properties to stimulation by an electric field which causes a charge separation or polarization in their atomic or molecular structures. Their polarization voltage decay time is short. In contrast, induced polarization is mainly a current-induced phenomenon resulting from several interrelated electrochemical mechanisms occurring on a larger scale with longer decay times. A physical or electrical model of induced polarization is not as simple as that of an ordinary dielectric. Although one can and should be as exact as is practical in measuring the IP effect, one is continually reminded of the heterogeneous nature of rocks and the macroscopic scale in which the IP phenomenon is created.

The important phenomena in rocks that cause induced polarization are diffusion of ions next to metallic minerals and, to a lesser extent nonmetallic minerals, electrochemical oxidation—reduction reactions between electrolytes and electron-conducting minerals, and ion mobility within the pore-filling electrolyte.

IP response

In theory, induced polarization is a dimensionless quantity, but in practice it is measured as a changing voltage with time or frequency. The time and frequency IP methods are fundamentally similar and only differ in the way of viewing and measuring waveforms. The inducing-current waveforms of the two methods are not the same, although this point of comparison alone is not significant beyond noting that frequency IP measurements are usually made while current is on and time IP measurements are made while the inducing current is off.

Consider the voltage-versus-time waveform shown on Fig.1.4. This is due to a pulsed current input square wave giving rise to charge and decay curves. The cycling period is long compared to the decay rate and the curve shapes are logarithmic with time. Due to the linear-system properties of earth materials, the frequency-measured IP voltage buildup of the charging cycle is a symmetric mirror image of the time-measured IP decay curve. The maximum charging cycle voltage amplitude of Figs.1.1 and 1.4 is proportional to the dc resistivity of the material being studied. A basic response parameter measured in the IP method is simply the amount of change of voltage ΔV or resistivity $\Delta\rho$ seen as a function of either time or frequency.

The frequency f of any particular periodic function of period T is given by:

$$f = 1/T \tag{1}$$

Referring to Figs.1.2 and 1.4, note that as frequency f is greater, time t of a proportional waveform interval or fraction of an interval is smaller. This

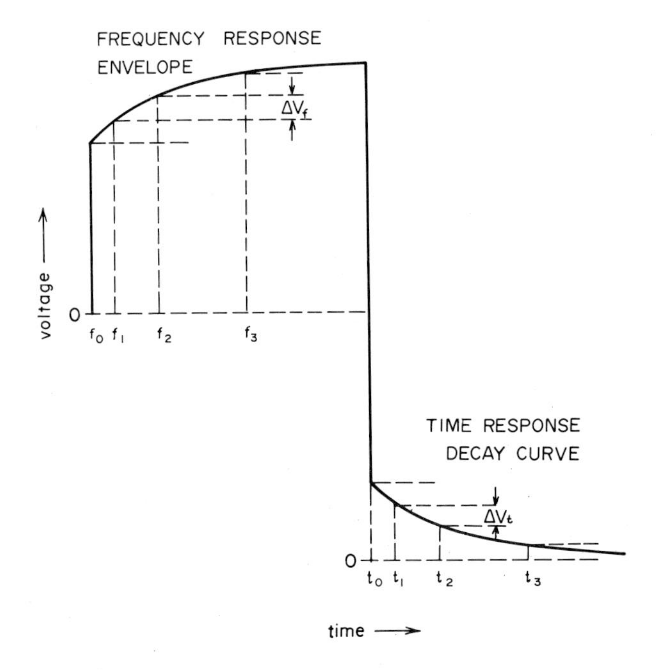

Fig.1.4. An IP charge and decay curve, using a long-period square wave. The logarithmic decay curve displays decay voltages at geometrically spaced times t_1, t_2, and t_3. The charge curve is the inverse of the voltage decay with geometrically spaced frequencies f_1, f_2, and f_3.

inverse proportional relationship between frequency and time also holds for fractions of periods of different waveforms, which can be written:

$$f \propto 1/t \tag{2}$$

Logarithmic voltage amplitudes, varying with both time and frequency, are indicated on Fig.1.4, and the conventional frequency-method waveform is on Fig.1.2. The inversion principle of eq. 2 applies to any time-to-frequency comparison, and only the time intervals need be proportionally adjusted when comparing different waveforms.

IP voltage responses for any given linear material are equivalent with time and frequency (Fig.1.4):

$$\Delta V_t \Big|_n = \Delta V_f \Big|_n \tag{3}$$

if there are logarithmically varying voltages measured over equivalently timed n spacings. This statement can be verified by taking derivatives of logarithms of normal or inverse functions.

The decay curve

IP decay curves (Figs.1.4 and 1.5a) are a cumulation of ΔV response intervals, and they provide a fundamental representation of the IP effect. Because of electrical instrument considerations, time-domain IP measurements in the field are often made by integrating the area under the decay curve, which gives an approximation of the total IP response. This kind of measurement will be discussed later in more detail.

Decay curves are observed to be generally logarithmic in form as has been noted by Wait (1958) and by others. To a first approximation, all IP decay curves can be treated as being logarithmic, and this fact can be used as a basis for defining the basic IP measurement. It is also necessary to assume that the time constant for all decay curves has a fairly uniform value. For a given decay curve the logarithmic voltage dependence can be written:

$$V_t = K_s - K_t \log t \qquad (4)$$

where K_s and K_t are constants over the IP range of interest. Note that K_s is a current-off voltage intercept and that K_t is the slope of the IP response decay curve. Though K_s is perhaps more fundamental from a theoretical standpoint it is more difficult to define and to measure precisely and carries less information than K_t.

With the frequency curve of Fig.1.4 or Fig.1.5b, an equation:

$$V_f = K_p - K_f \log f$$

can be written similar to eq. 4. Here K_p is a current-on voltage intercept and because charge and decay can be taken as being logarithmic, $K_t = K_f$.

The geometrically increasing time instants t_1, t_2, etc., shown in Fig.1.4,

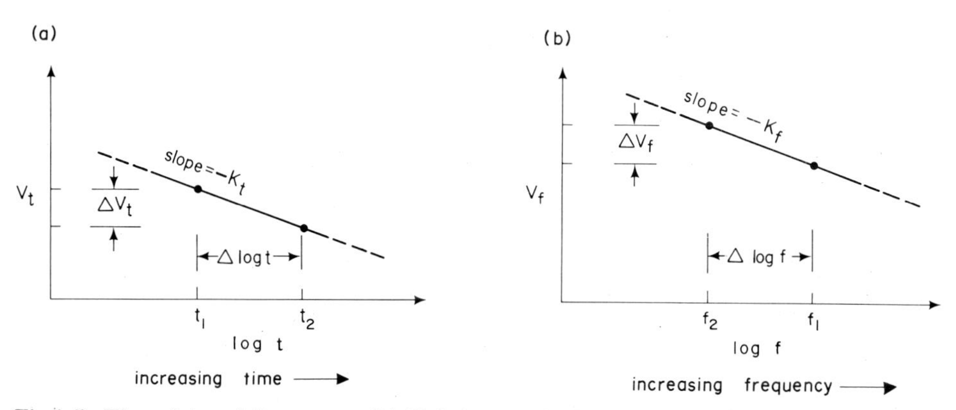

Fig.1.5. Time (a) and frequency (b) IP behavior shown in the IP response region where equal voltage increments ΔV_t and ΔV_f are functions of logarithmic time or frequency. The slope of the curves, equal to $-K_t$ of eq. 4, is the measure of IP response.

correspond with equally spaced decreasing voltages V_1, V_2, etc., in a predictable way. The decay-voltage amplitude interval ΔV_t is a function of progressing time, and it is this voltage interval that must be measured against time.

The total IP decay-voltage response in the time domain would be:

$$\sum_n \Delta V_t = V_s \tag{5}$$

and with longer time increments, $\Delta V_t \rightarrow V_s$. From Fig.1.2, the incremental response interval in the frequency domain is:

$$\Delta V_f = V_2 - V_1$$

IP MEASUREMENT

If IP measurements are to be standardized and compared, ΔV needs to be described more exactly than merely as a voltage change with time or frequency. To accomplish this refinement requires a close look at IP response characteristics. The ΔV interval should be precisely defined in order to be a proper quantitative measure of IP response. Let us first examine the slope K_t of the IP response curve as given in eq. 4 and shown in Fig.1.5.

For a given logarithmic decay curve, the constant voltage interval ΔV_t taken over a logarithmic time interval is proportional to the decay-curve slope K_t of eq. 4 and Fig.1.5, and:

$$\Delta V_t = -K_t \Delta \log t \tag{6}$$

the voltage V_t decreasing with increasing time. A similar relationship holds with frequency:

$$\Delta V_f = -K_f \Delta \log f$$

Figure 1.5 diagrammatically displays logarithmic response curves in time and in frequency showing increments of ΔV_t and ΔV_f. If $\Delta \log t$ (or $\Delta \log f$) is equal to one, from eq. 6, $\Delta V_t = -K_t$. The slope of the IP response curves shown in Fig.1.5 and as given in eq. 6 becomes the important measure of the IP phenomenon.

For consistency's sake in field surveying practice, the decay curve or IP frequency function is sampled at certain standardized frequency points or over specified time intervals. If this time interval or frequency ratio is made a factor of ten, since $\log 10 = 1$, ΔV_t and ΔV_f give the slope of $-K_t$ and $-K_f$ directly.

The basic IP measurement $\Delta V/V$

To obtain a normalized measure of the IP effect, the ΔV IP response value given in eqs. 3, 5 and 6 must be divided by the primary voltage V_p because ΔV is proportional to and linear with the stimulating primary voltage. IP response ΔV or $\Delta \rho$ is observed as a voltage or resistivity change. In order to form an independent IP parameter, the IP response must be formulated as a ratio with respect to voltage V or resistivity ρ. The fundamental basic IP function can then be written as $\Delta V/V$ or $\Delta \rho/\rho$. The primary IP parameter evaluated with time, chargeability M, can be written as:

$$M = \Delta \rho_t / \rho \qquad (7)$$

and if measured with dependence on inducing frequency, the IP frequency effect (FE) is:

$$FE = \Delta \rho_f / \rho \qquad (8)$$

The time and frequency intervals over which $\Delta \rho$ is measured in eqs. 7 and 8 are usually standardized to logarithmic unity in order to make IP measurements self-consistent.

Compared to the wide variation in the electrical conductivity of minerals and rocks, the IP response of earth materials is not large. Sensitive measurement techniques are necessary. And because of the sensitivity considerations of different types of field instruments, there are several different ways of measuring and expressing both chargeability and frequency effects. The quantities given in eqs. 7 and 8 are small fractions and are therefore usually multiplied by constants from 100 to 1,000 in order to put the primary polarization parameter into the range of commonly used numbers.

Phase-angle IP measurement

Two sinusoidally varying waveforms with the same oscillatory frequency can be compared by noting their shift with time or angular fraction of a wavelength. The waveforms shown in Figs.1.1 and 1.2 can be redrawn as pure sine waves (Fig.1.6a) to illustrate the phase-lag angle between a polarization voltage signal and the stimulating current reference waveform.

An IP phase-lag angle can be plotted on a diagram similar to the time and frequency spectral plots of Fig.1.5, as seen on Fig.1.6b. If the polarization decay curve is logarithmic in shape, the IP phase-lag angle is constant over the range of interest, yielding the horizontal line of Fig.1.6b. It has been shown by Zonge et al. (1972) that IP phase lag can be equated to frequency effect and chargeability.

IP phase lag can be separated into two components with respect to the current waveform, as shown on Fig.1.7. The in-phase and out-of-phase components are more readily measured with modern electronic systems than the phase angle alone, and the two components provide additional useful information about IP phenomena. The in-phase and out-of-phase components can also be shown as rotating vectors, as illustrated in Fig.1.8, with the phase-lag

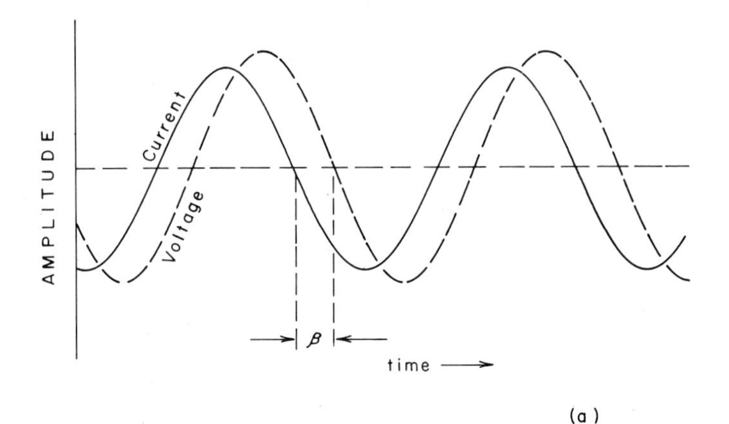

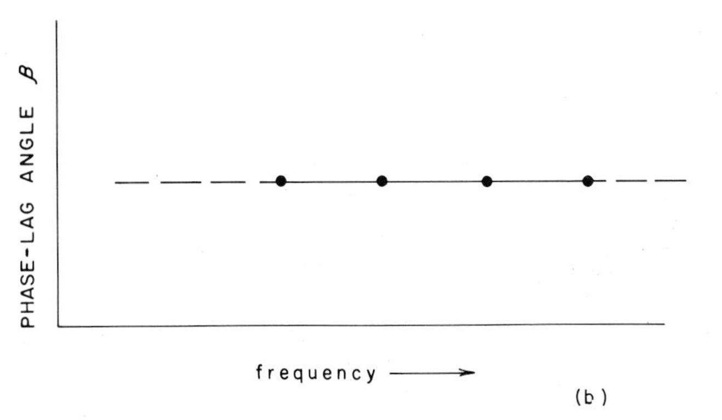

Fig.1.6. IP phase determinations. (a) phase lag β between input (solid) and output (dashed) sinusoidal waveforms, and (b) ideal IP phase spectrum diagram.

angle β between the current and voltage vectors. Thus, the phase angle can be defined as the angle whose tangent is the ratio between the imaginary and real components, or:

$$\beta = \tan^{-1} \frac{V_{\text{imag}}}{V_{\text{real}}}$$

where V is voltage.

Errors

The error in an IP value is dependent on how accurately (actually how inaccurately) ΔV and V are measured. Here standard error analysis techniques apply, and the total percent error in an IP reading would be:

$$\text{total percent error} = \left| \frac{\delta \Delta V}{\Delta V} \right| 100 + \left| \frac{\delta V}{V} \right| 100$$

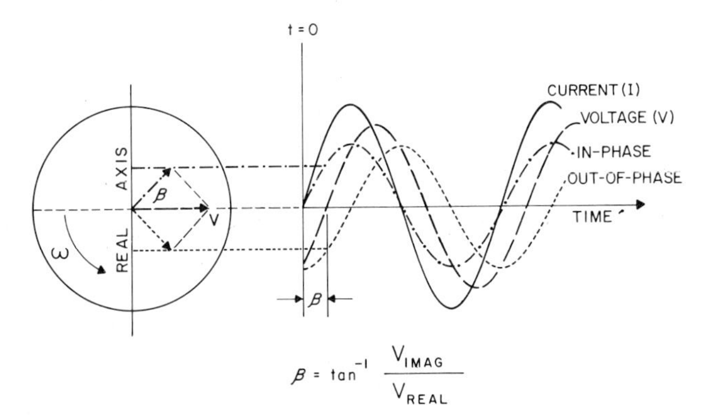

Fig.1.7. Components of the IP phase-lag angle.

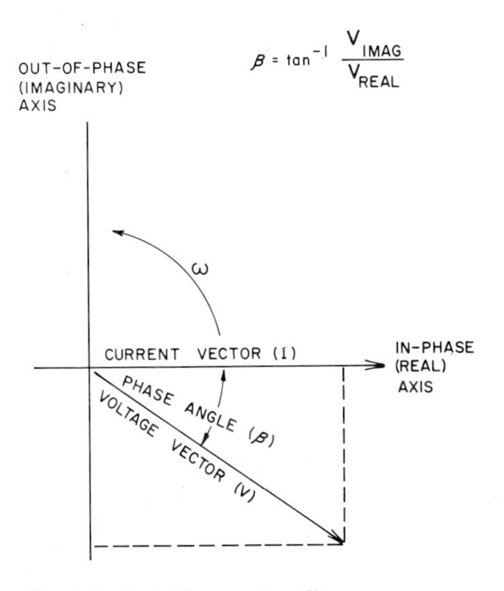

Fig.1.8. Rotating vector diagram.

where δ symbolizes the relative error or average deviation of the quantity. The probable error in ΔV is much greater than in V, and an expression for this error can be derived. ΔV is a small quantity and represents the difference between two voltage values, so its error is at least double the error of V. In the frequency method, an error in inducing current measurement also creates an error in voltage values. Relative voltages to obtain ΔV are taken at different frequency or time intervals, and if this interval is large, the error in ΔV is smaller. Absolute IP voltage values are not as important as consistent relatively accurate values because relative errors can be cancelled or calibrated.

There are several ways of measuring chargeability and frequency effect, depending on the purpose and object of an investigation, the equipment available, and geological conditions. These methods will be treated later after developing fundamentals of electrical prospecting.

Units used in IP surveying

There is a continuing tendency for the world's engineers and physical scientists to turn to the rationalized mks (meter-kilogram-second) system of units for basic definitions of measurements and terms. This trend is brought on by the pressing needs for mutual understanding and the benefits and convenience of universal communication.

For the sake of uniformity it is recommended that ohm-meters be used for units in resistivity and that when presenting IP data, values of frequency effect or chargeability be given in standard units that can be translated to other systems. Frequencies or cycling times and off-time duration must be made clear in order to give deserving work enduring qualities.

THE GLOSSARY

One of the important factors in the communication of information is the precise meaning of the various technical terms that are employed in describing a body of knowledge. Induced polarization is a recently developed, rapidly growing area of geophysical study and has borrowed heavily for its vocabulary from such diverse disciplines as electrochemistry, physics, electronics, geology, and mathematics. The technical languages of these different fields are not always compatible, and there have been occasional misunderstandings of meanings. In addition, a jargon has developed within different groups working on the same applied research problems, and meanings have sometimes been hazy and imprecise. For these reasons, the writer has compiled a glossary of the terms used in induced-polarization and resistivity prospecting. The glossary is placed at the back of this book where the reader may find it convenient for reference. Whenever possible, existing glossaries have been relied upon, such as the *Encyclopedic Dictionary of Exploration Geophysics* by R.E. Sheriff (1973) and *A Dictionary of Electronics* by S. Handel (1966). Usages that have been introduced, defined, and have been generally accepted in geophysical literature are noted. However, most of the terms given in this glossary have evolved through ordinary laboratory and field usage, and the writer is deeply grateful to several geophysicists who have contributed corrections and suggested changes and additions to this list.

Definitions of terms used in resistivity surveying are included in this glossary when the term is fundamental or is also used in IP exploration.

Chapter 2

RESISTIVITY PRINCIPLES

Before embarking on a discussion of the IP method, it is appropriate to review the principles of resistivity exploration, because induced polarization is fundamentally based on the nearly direct current, electrical resistivity behavior of rocks. This is not to say that induced polarization is just a refinement of resistivity anymore than to imply that alternating current is a special case of direct current; it is the other way around. However, just as an understanding of dc resistance is needed before studying impedance, it is necessary to know something of dc resistivity before coping with rock impedance and the IP effect.

Certainly a good half of the information obtained in an IP survey is in the form of resistivity measurements. In the interpretation of IP data, IP values, in the form of frequency effect or chargeability, are frequently compared with their resistivity counterpart. Resistivity maps are useful for correlation with subsurface rock type and structure, and data can be interpreted to give basic electrical property information in three dimensions.

The ratio of an IP value to resistivity is the prime term of the *metal factor*, a resistivity-related parameter which sometimes can be useful in interpretation. Resistivity is fundamental in the meaning of this ratio. In a similar way, resistivity is basic to the quantitative interpretation of IP data inasmuch as resistivity contrast information is necessary before attempting to match type IP curves.

The resistivity method and the dc electrical response of earth materials are an extensive subject in themselves that has been thoroughly treated in several recent excellent books. Van Nostrand and Cook (1966) have published a broad treatment in their *Interpretation of Resistivity Data* (USGS Prof. Paper 499). Kunetz (1966) has rewritten in good form the Schlumberger and Compagnie Générale de Géophysique in-house manuals as *Principles of Direct Current Resistivity Prospecting*. Ward (1967) and his colleagues have presented a treatise on the electromagnetic methods of exploration in the Society of Exploration Geophysicists' *Mining Geophysics*, Vol. II, providing a summary of higher frequency electrical methods. Keller and Frischknecht's (1966) fine book, *Electrical Methods in Geophysical Prospecting*, also contains information on the subject, as does Griffiths and King's (1965) *Applied Geophysics for Engineers and Geologists*. A physically appealing treatment is made in Grant and West's (1965) *Interpretation Theory in Applied Geophysics* and in the intermediate-level book, *Direct Current Geoelectric Sounding*, by Bhattacharya and Patra (1968).

RESISTIVITY THEORY

This section is intended to refresh the reader in some of the fundamental principles of electricity, especially as applied to three-dimensional materials. It is also desirable to develop a foundation for discussing near-dc electrical geophysical exploration methods.

Ohm's law

Ohm's law is usually given as:

$$V = IR \tag{9}$$

where V is potential difference in volts, I is electrical current in amperes, and R is resistance in ohms. This equation expresses the proportional relationship between current in an electrical circuit and the electromotive force that is necessary to drive and maintain it. It defines resistance as a ratio factor of voltage to current, and if the ratio is constant over a fairly large voltage range, the circuit behavior is described as 'linear' or 'ohmic'.

Equation 9 does not imply anything about the dimensions of substances or the nature of materials involved in the electrical measurement. So, to describe the electrical resistance of a one-dimensional material, such as a length L of wire, the symbol ρ_l is used for the linear resistivity of the wire and:

$$\rho_l = \frac{V}{I}\left(\frac{1}{L}\right) \tag{10}$$

where ρ_l has the dimension of ohms per unit length. The measurement is diagrammed as Fig.2.1.

In the case of a two-dimensional conducting material, such as an elec-

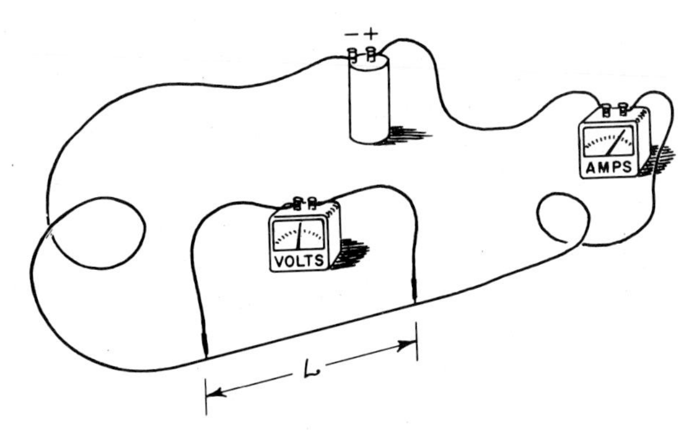

Fig.2.1. Linear resistivity measurement of a one-dimensional material.

trically conducting sheet of recording paper, the areal resistivity of the sheet ρ_s could be measured by passing current between line electrodes of length w at parallel edges of a rectangle of the material. The potential difference V between the lineal current contacts (or electrodes) spaced a distance L apart would be given in the form shown on Fig.2.2 or as:

$$\rho_s = \frac{V}{I}\left(\frac{w}{L}\right) \tag{11}$$

and the units of the areal resistivity of the sheet would be ohms per square, or simply ohms. Note that the measurement will still need to involve only two contact electrodes if they are parallel and linear, but that four electrodes could be used at the corners of a rectangle, as in Fig.2.3, if the electrode contact size is small compared to the distance between them. The ratio between w and L is the important quantity; therefore, the actual spacing interval between the four electrodes is unimportant as long as the electrodes are small and the w/L ratio is maintained.

The physical treatment of two-dimensional resistivity has practical importance in the interpretation of data from linear subsurface bodies. It is also used in potential field and resistivity modeling where the response of two-dimensional subsurface bodies is simulated by using conductive paper with better conducting shapes or insulating shapes on it.

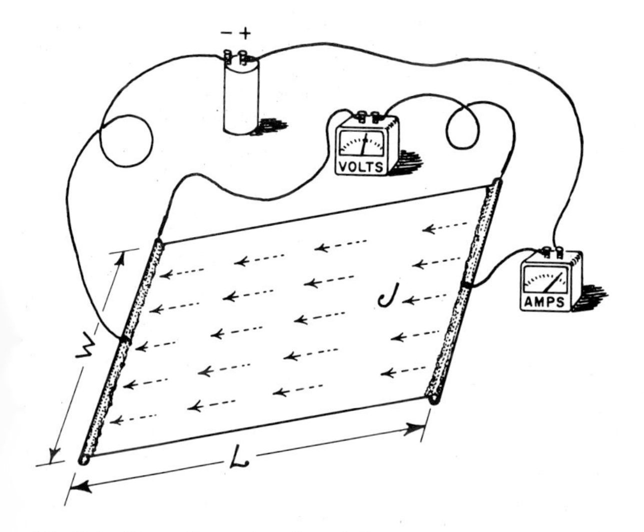

Fig.2.2. Two-dimensional resistivity measurement using two contact electrodes on the conductive sheet.

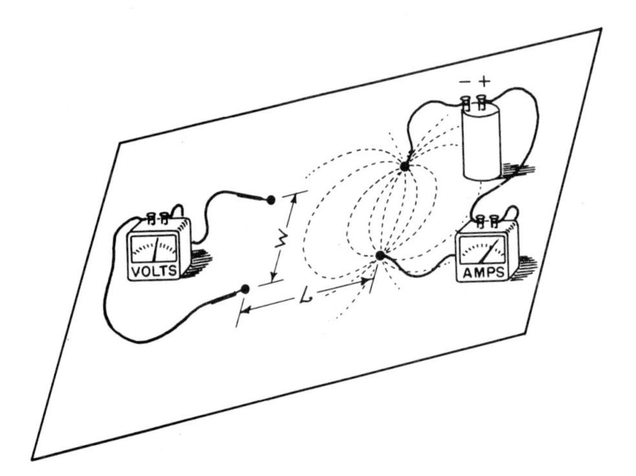

Fig.2.3. A two-dimensional resistivity measurement using two pairs of small contact electrodes.

Three-dimensional resistivity

If a current is passed between opposite faces of a resistive cube of side length L, as shown in Fig.2.4, the formula for its three-dimensional resistivity is:

$$\rho = \frac{V}{I}\left(\frac{L^2}{L}\right) \tag{12}$$

and ρ has the dimensions of ohms times length. In isotropic materials, conductivity (σ), measured in mhos per meter, is simply the inverse of resistivity: $\sigma = 1/\rho$. It would be possible to derive eq. 12 from the diagram shown as Fig.2.5, which starts with the vector relationship between electric field intensity $\overleftarrow{E}$ and current density $\overleftarrow{J}$ as:

$$\overleftarrow{E} = \rho\overleftarrow{J} \tag{13}$$

In this equation, $\overleftarrow{J} = I/L^2$ has dimensions of amperes per unit area. Equation 13 is actually a basic expression of Ohm's law for materials with one, two, or three dimensions. It could have been used to derive eqs. 10, 11 and 12 by using the proper units for ρ and by appropriately defining current density $\overleftarrow{J}$ in amperes, amperes per length, or amperes per area, respectively. The units of electric field $\overleftarrow{E}$ in any of these dimensions are volts per length. E can be replaced by $-\mathrm{d}V/\mathrm{d}L$ because conventionally potential decreases in the direction of the electric field and the current flow. Then by looking at the small potential difference $\mathrm{d}V$ over the small length $\mathrm{d}L$:

$$\mathrm{d}V = -\rho\,\frac{I}{L^2}\,\mathrm{d}L \tag{14}$$

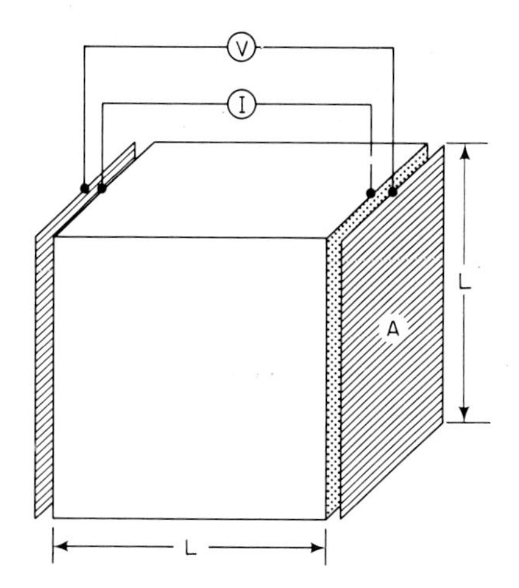

Fig.2.4. Resistivity measurement of a cube of side length L using two current and potential measuring electrodes.

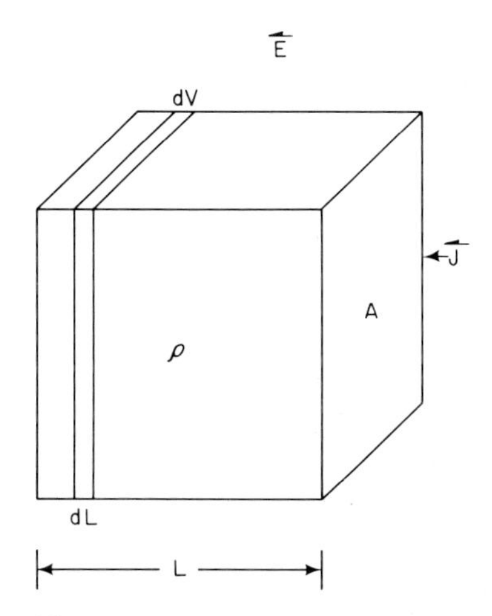

Fig.2.5. Illustration of a way of deriving Ohm's law in three dimensions from a cube of resistivity ρ in an electric field E.

which by integration:

$$\int_0^V \mathrm{d}V = -\rho I \int_0^L \frac{\mathrm{d}L}{L^2}$$

would also give eq. 12.

Considering a point electrode on the surface of an electrically conducting solid, as shown on Fig.2.6, eq. 14 becomes:

$$\mathrm{d}V = -\rho J\,\mathrm{d}r \tag{15}$$

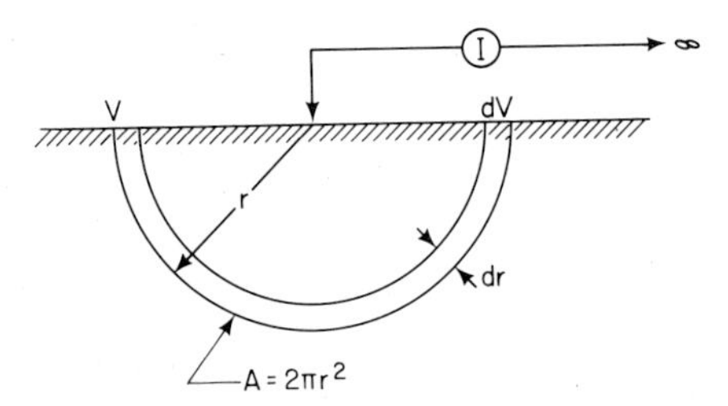

Fig.2.6. Section through point current electrode on surface of a conducting solid, showing how potential V is related to resistivity ρ, current I, and distance from electrode.

In this equation, $J = I/2\pi r^2$, and:

$$dV = -\rho I \frac{dr}{2\pi r^2} \tag{16}$$

It follows by integration of both sides of the equation that:

$$V = \frac{\rho I}{2\pi r} \tag{17}$$

Equation 17 is important because it describes the electrical potential about a point contact electrode on the surface of a three-dimensional material of resistivity ρ. This relationship is also useful in deriving expressions for the geometric factors of the various electrode arrays, as will be investigated later.

Resistivity and conductivity

There is a basic physical difference between resistivity and conductivity, even though in isotropic materials one term is merely the reciprocal of the other. Resistivity is a measure of the opposition to flow of charge in a material, whereas electrical conductivity is the flow mobility of charge carriers. Conductivity σ is derived from the relationship $\sigma = ne\mu$, where n is the density of charge carriers, e is their charge, and μ is their mobility measured by velocity in meters per second per unit electric field. The charge carriers may be ions, electrons, or holes (the absence of a charge).

Although the concept of conductivity is perhaps physically more fundamental than that of resistivity, for the sake of consistency, in this book resistivity will usually be referred to rather than conductivity because of the prior usage in the study of resistivity as a geophysical exploration method. Low-frequency IP measurements are more often related to Ohm's law and notions of resistivity. The more complex higher frequency electrical phenomena usually introduce the concepts of conductivity. Other factors remaining constant, apparent resistivity is proportional to potential difference and resistivity is directly related to voltage values measured in the field.

The physical units or dimensions of resistivity in the mks system are ohm-meters (Ω m), which is resistance times length. Equations that define the resistivity of materials by measurement, such as are developed in this chapter, can be rearranged as $\rho = (V/I)K$, where K is known as a geometric factor or form factor with units of length. The geometric factor K is constant for a given array and spacing, varying with the electrode interval.

Some groups doing resistivity and IP surveying have preferred to leave the 2π term from eq. 17 on the same side of the resistivity equation with ρ, thus creating a $\rho/2\pi$ unit of resistivity, the definition for which is 2π or 6.283 times the unit resistivity of ρ. Table II gives the relationship of the various units that have been used for resistivity measurements. It is noteworthy to see that to within 5%, one $\rho/2\pi$ ohm-feet equals two ohm-meters.

TABLE II

Relationships between units used in resistivity surveying

	ohm-centimeters	ohm-feet	ohm-meters	ohm-feet ($\rho/2\pi$)
ohm-centimeter	1.0	0.0328	0.01	0.00522
ohm-foot	20.48	1.0	0.305	0.159
ohm-meter	100	3.28	1.0	0.522
ohm-foot ($\rho/2\pi$)	191.5	6.283	1.915	1.0

Potential field of a current electrode pair

If two current contact electrodes of source strength $+I$ and $-I$ are placed on the planar surface of a conducting solid at positions A and B, as shown on Fig.2.7, then by combining (superposing) potentials from eq. 17 at any point M on the solid, the resulting potential V_M is:

$$V_M = V_A - V_B = \frac{\rho I}{2\pi}\left(\frac{1}{r_A} - \frac{1}{r_B}\right) \tag{18}$$

The potential difference V between any two potential sampling points M and N (Fig.2.8) on the solid is:

$$V = V_M - V_N = \frac{\rho I}{2\pi}\left(\frac{1}{r_{AM}} - \frac{1}{r_{BM}} - \frac{1}{r_{AN}} + \frac{1}{r_{BN}}\right) \tag{19}$$

This is the general equation for *any* four-electrode resistivity measuring configuration on a plane surface.

Sometimes the question is asked why it is necessary to have four separate contacts with the ground in order to make a resistivity measurement, when an ordinary ohmmeter needs only two contacts in measuring resistance. The answer is that if only two contacts are used in finding the electrical resistivity of a solid, the resistance of the contacts at the two points must be

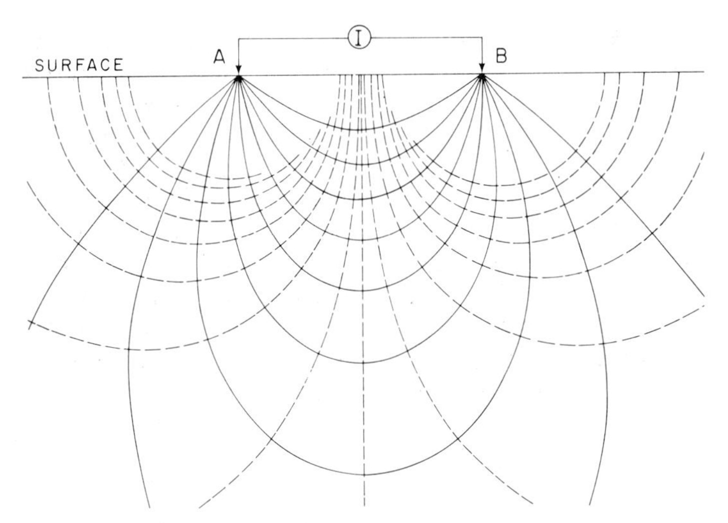

Fig.2.7. Potential surface distribution and current flow lines in a conducting solid about two current contact electrodes.

known and these may be large. To find the resistance of a point contact, the exact shape of the contact and its relationship with the measured homogeneous material must be known. Also, there is a difficulty in making electrical contact using single electrodes for both current and voltage measurements because the electrical properties of the ground may become temporarily disturbed by locally large current densities. To avoid these basic problems, two separate electrical circuits with two contact electrodes in each circuit are used, one pair connected to a transmitter (current sender) at *A* and *B* and one pair connected to a receiver (potential sensor) with contacts at *M* and *N* (Fig.2.8). With a four-electrode system, if the size of contacts is much smaller than the distance between them, the size and resistance of the contacts become unimportant and the disturbed-ground problem is avoided.

Referring to Fig.2.7, an electric field is created around two current electrode locations. The potential electrode pair does not disturb this field and only acts to sample the electric potential present by virtue of the current field.

When the pair of small-sized potential measuring points *M* and *N* are between *A* and *B* (Fig.2.8), the potential field is quite simple and the equipotential surfaces in this region are almost parallel. This electric field is sometimes called a *parallel field*, or a *plane-wave field.* There are several conventional electrode arrays used for resistivity surveying in which the dipolar, potential gradient measuring electrode pair measures an approximate parallel electric field. These field arrays include the *AB* rectangle array, gradient array, Schlumberger array, and to some extent, the Wenner array. The appropriate formulas for these arrays can be readily derived from eq. 19,

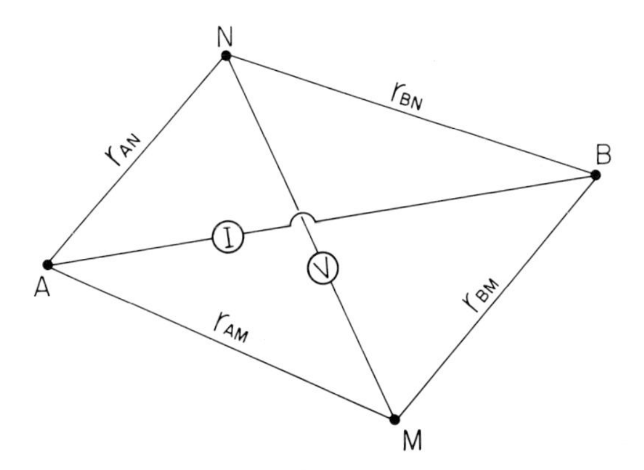

Fig.2.8. Generalized current and potential electrode arrangement using four point contacts on surface of a solid.

the solution of which gives the form factor K of the particular array. In the absence of a subsurface conductor, all the parallel field conditions tend to have $\Delta V/\Delta a = C$, that is to say, the rate of change of potential with the interelectrode distance is nearly constant. The reader is urged to see for himself that the formulas for these arrays are readily derived from eq. 19. Figure 2.9 is a contour plan map of the form factor K_{AB} of the gradient array which was derived from eq. 19.

If the potential gradient measuring electrode pair is much nearer to one of the current electrodes than to the other, the potential differences about that point electrode are measured. In this event, r_{BM} and r_{BN} are very large and eq. 19 becomes:

$$V = \frac{\rho I}{2\pi}\left(\frac{1}{r_{AM}} - \frac{1}{r_{AN}}\right) \tag{20}$$

and V approximately varies as $1/r$. Arrays that employ this field relationship are the three-array and the pole—dipole array. The potential-drop-ratio (PDR) method is also concerned with the field of a single current electrode, comparing two adjacent potential field gradients in a simple ratio. The commonly used electrode arrays for resistivity surveying are shown in Fig.2.10. The type of array should be chosen on the basis of equipment available and type of sought-for subsurface feature, but often the array in use is a matter of personal habit.

In investigations of the potential field about a point current electrode by means of a single potential electrode test probe in which the companion current and potential electrodes are a great distance away, the configuration is called the pole—pole array. This two-electrode array has also been called the 'half-Wenner' array by Keller and Frischknecht (1966), and the geometric factor is the same as for the Wenner array.

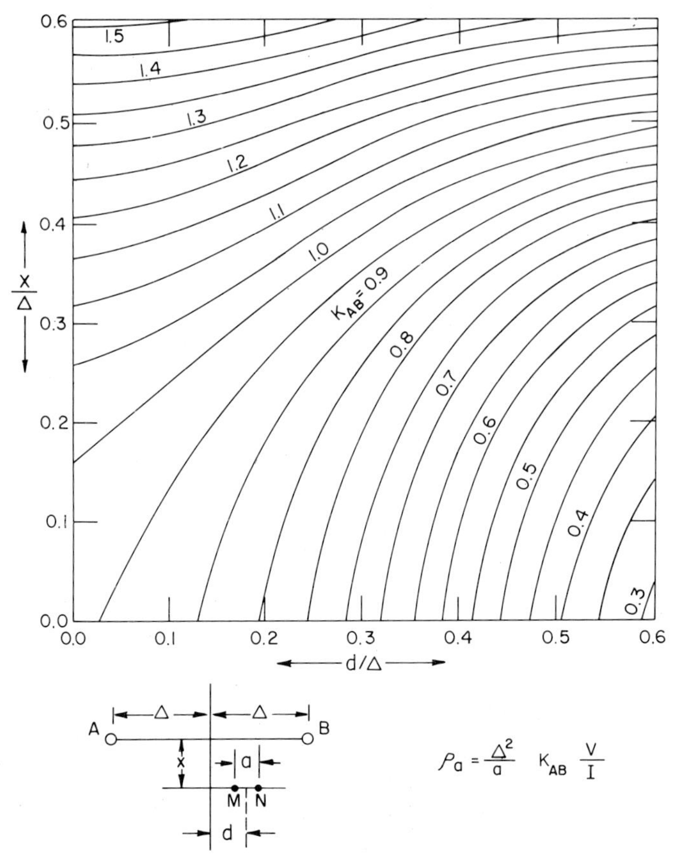

Fig.2.9. Plan map of gradient array, showing contours of the form factor K_{AB}.

If the dipole potential gradient measuring electrode pair is farther from the current pair than the distance between pairs, eq. 19 will show that approximately the potential gradient of a dipolar electric field is being measured and that V will vary as $1/r^2$. The dipole—dipole electrode array uses this relationship in its measurements. Electrode pairs, as used in this method, are usually kept on parallel line segments, either on the same survey line or on adjacent parallel traverses. The dipole—dipole array measures a higher order potential field, but it requires a larger current in order to excite a voltage comparable to that measured using other arrays.

There are therefore three basic types of electric field measurements that can be made near a current electrode pair, as shown in Fig.2.11 and summarized in Table III. This tabulated classification is meant to show only the comparative relationships between the most common types of arrays used in IP surveying. Generally, the higher-order, dipole source array gives a higher resolution of lateral resistivity variations than the plane-wave arrays, and this

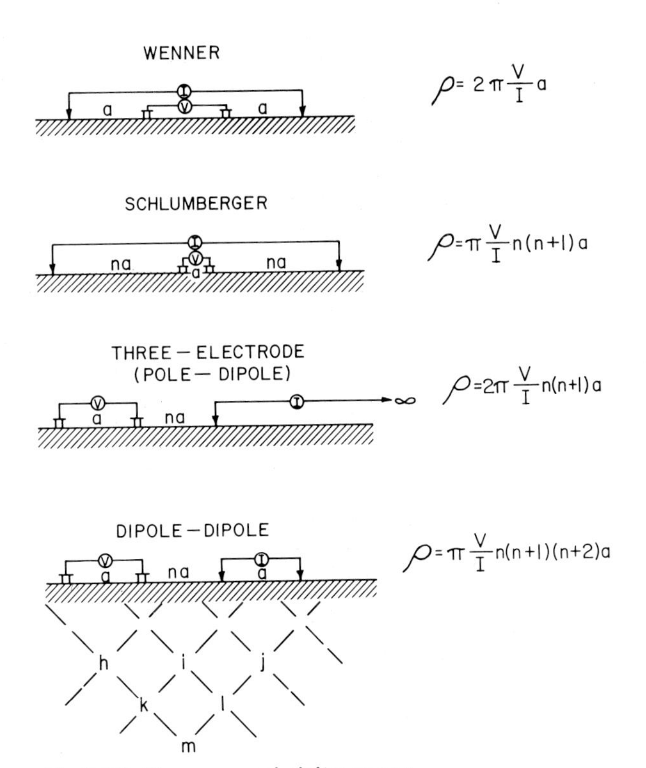

Fig.2.10. Common resistivity arrays.

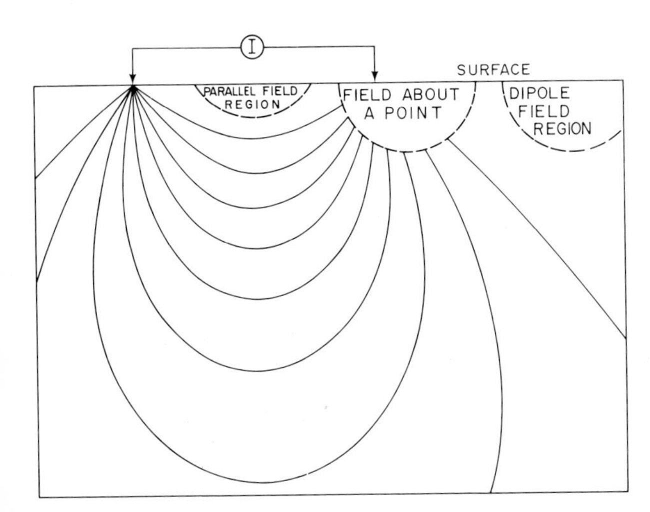

Fig.2.11. The three basic regions about two current electrodes where potential measurements are made.

TABLE III

The common electrode arrays, classified according to type of potential field being measured

Type of potential field	Electrode array	Apparent resistivity formulas
Plane wave or parallel field $V \propto a$	*AB* rectangle array gradient array	$\rho_a = \frac{\Delta^2}{a} K_{AB} \frac{V}{I}$
	Schlumberger array	$\rho_a = \pi \frac{V}{I} n(n+1)a$
	Wenner array	$\rho_a = 2\pi \frac{V}{I} a$
Potential field about a single current electrode $V \propto 1/r$	three array	$\rho_a = 4\pi \frac{V}{I} a$
	pole—dipole array	$\rho_a = 2\pi \frac{V}{I} (n+1)an$
	PDR array	$*PDR = \frac{V_{MN}}{V_{RS}} \left[\frac{V_{RS}}{V_{MN}}\right]_{\text{normal}}$
Potential field about a current dipole pair $V \propto 1/r^2$	dipole—dipole (in line)	$\rho_a = \pi \frac{V}{I} n(n+1)(n+2)a$

*This is a pure ratio rather than an apparent resistivity formula.

fact is sometimes desirable in mineral exploration surveys. Figure 2.12 shows comparative resistivity profiles over a buried sphere.

All of the preceding discussion concerning types of electric fields and electrode arrays has assumed that the earth's resistivity really is homogeneous and isotropic. Since these ideal conditions are seldom actually encountered, the resistivity measured in field situations is a function of complex subsurface conditions and is therefore an *apparent resistivity*.

Refraction of current flow lines

When an electric current line crosses a plane boundary between materials of contrasting resistivities, it is refracted in much the same way as is a light ray passing between optical media. The electric field and potential are continuous across the boundary, so that the parallel components must be equal, thus:

$$\rho_1 \vec{J}_{\parallel 1} \equiv \rho_2 \vec{J}_{\parallel 2} \tag{21}$$

The normal component of current flow is the same on both sides, therefore:

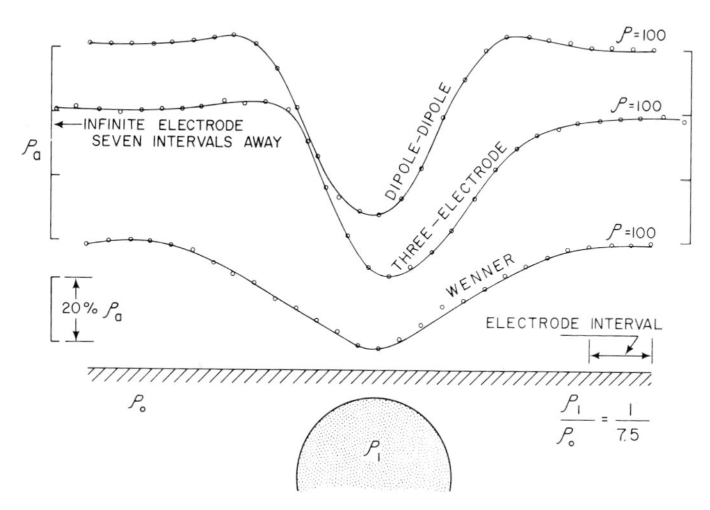

Fig.2.12. Resistivity profiles over a buried sphere, giving a comparison of the resolution of three commonly used electrode arrays.

$$\vec{J}_{\perp 1} = \vec{J}_{\perp 2} \tag{22}$$

that is, there is no accumulation of charge at the boundary of a purely resistive body. The solution of eqs. 21 and 22 shows that current is indeed refracted at the boundary, the tangents of the angles of incidence and refraction being inversely proportional to the resistivities. Thus:

$$\frac{\tan \theta_1}{\tan \theta_2} = \frac{\rho_2}{\rho_1} \tag{23}$$

and a current line penetrating into a more resistant medium is bent toward the normal, as shown on Fig.2.13.

Equipotential surfaces

Schlumberger (1920) has pointed out the value of considering equipotential surfaces in resistivity surveys, a technique stemming from what is called the equipotential survey method. Specifically, if a conductive body is in a parallel electric field, the current flow lines will tend to be deflected through the body, while the equipotential surfaces, being orthogonal to the current lines, will conform to the body, as shown on Fig.2.14.

The interpreted conditions shown in Fig.2.14 can be observed in field resistivity surveys, for example, when using the gradient array to map potential differences. Note that a fixed interval between a potential measuring electrode pair will show a larger voltage difference measurement just outside a conductive body, while within the conductive body the difference is smaller, as would be expected from the concept of apparent resistivity.

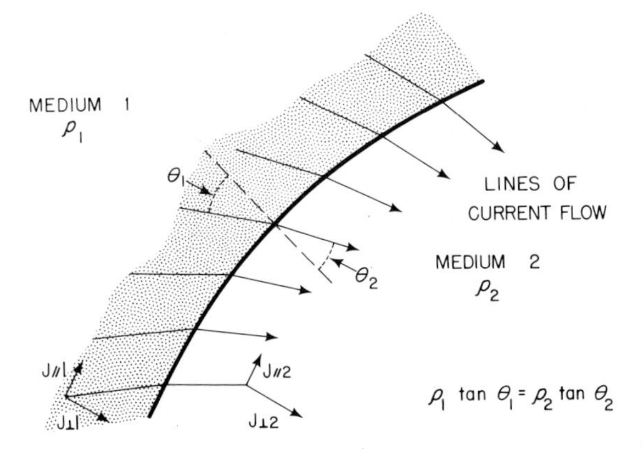

Fig.2.13. Refraction of a current flow line passing through an interface between contrasting resistivities.

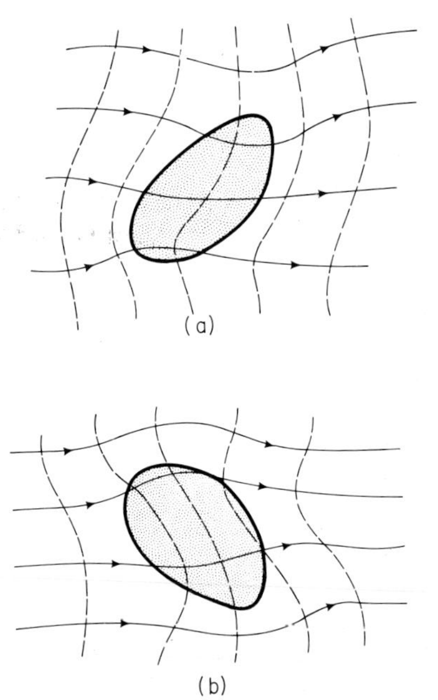

Fig.2.14. Current flow lines (solid) and equipotential surfaces (dashed). (a) About a conductive subsurface body. (b) About an insulating subsurface body.

Resistivity anisotropy

From observation of the successive equipotential lines on a planar surface through a point current source in an electrically anisotropic material, the lines are found to form a family of concentric ellipses. These are the surface intersections of the 'ellipsoid of anisotropy', and the ratio of major to minor axes is proportional to the ratio of the square roots of true resistivities, $\sqrt{\rho_T}$ and $\sqrt{\rho_L}$, in the transverse and longitudinal directions, as shown on Fig.2.15. The coefficient of anisotropy λ is given as:

$$\lambda = \sqrt{\rho_T / \rho_L} \tag{24}$$

and is always more than one but generally less than two.

Paradox of anisotropy

It may seem odd that the apparent resistivity measured in a direction normal to the bedding or foliation by a conventional array is always less than that measured parallel to the bedding. This effect is just opposite from the effect of bedding on the actual true resistivity value, and it results from electrical anisotropy and the geometric effect of the electrode array in the measurement. It can be shown that the measured apparent transverse resistivity will equal the true longitudinal resistivity of the material. This unusual behavior is called the *paradox of anisotropy*.

Superposition and reciprocity

The *principle of superposition* states that potentials and electric fields are algebraically additive. Of course, vector addition must be employed in the case of vector fields. This superposition concept is important in deriving geometric factors and type curves for electrical data.

The *principle of reciprocity* states that in any medium, whether anisotropic or heterogeneous, the potential at a point M due to a current source at A is the same as it is if current and potential positions were reversed. The principle holds for any electrode array. As a consequence of this principle, the locations of current and potential measuring pairs in any array can be

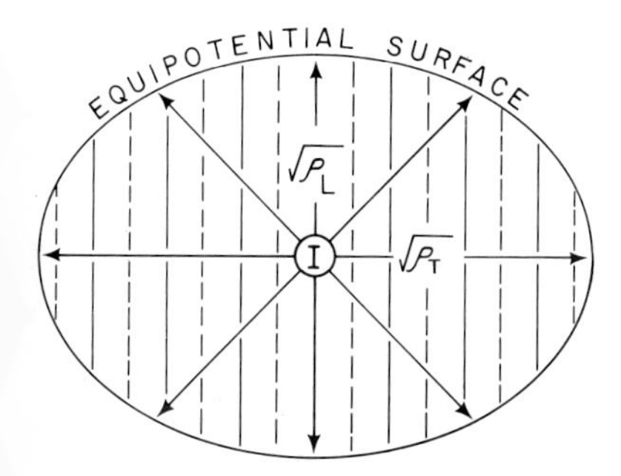

Fig.2.15. An 'ellipse of anisotropy' showing an equipotential surface and the transverse ρ_T and longitudinal ρ_L resistivities.

reversed and the measured values will be the same. Knowledge of this principle is useful in many field survey situations, for example, in avoiding the time-consuming placement of current electrodes in impractical places.

Resistivity type curves by the method of images

The usefulness of image theory in solving current and potential distribution problems can be illustrated in the example of an electric-field analysis between two equal static charges. Since a perpendicular conductive plane placed midway between the charges is on an equipotential surface, if one of the charges is removed, the field on the other side of the conductive plane remains the same.

If we visualize an electric charge of strength A near a conducting plane, as illustrated in Fig.2.16, we could replace the conducting sheet with an image charge of strength B, where B is a function of A, such as kA, k being a constant lying between $+1$ and -1. This is the image theory of electrostatics, and it can be appropriately modified to fit our needs.

Stating the electric-field problem in another way, the plane boundary between media of contrasting resistivities can be replaced by an image current source.

Now start again with a current source of strength A in a conducting material (medium 1) with a material of different resistivity (medium 2) nearby, as shown in Fig.2.17. A flat plane separates the two media. To an observer in medium 1, if medium 2 is removed and replaced by medium 1

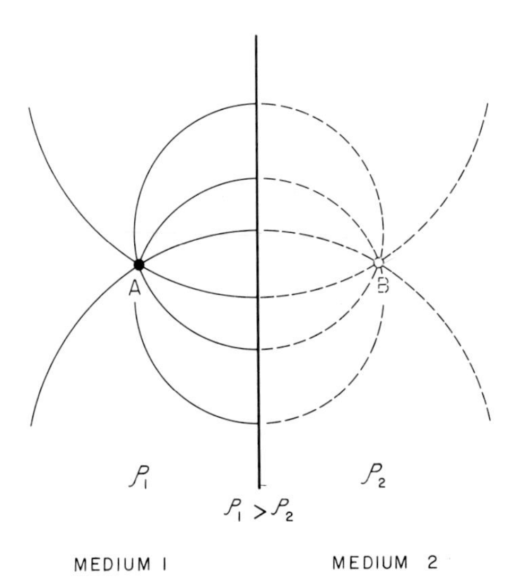

Fig.2.16. An electric charge A near a conducting plane and the electrical image B.

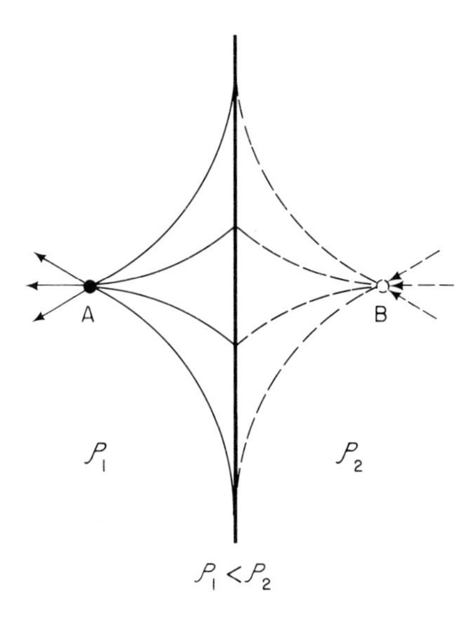

Fig.2.17. An electric current source of strength A in a conducting material near an interface with a contrasting resistivity.

and an image source equal to kA, the electric field in the original medium 1 would remain the same. At any point M, the potential is then:

$$V_M = \frac{A}{r_1} + \frac{kA}{r_2} \tag{25}$$

But by eq. 17, the potential of a current source is $V_M = I\rho_1/2\pi r$:

$$V_M = \frac{I\rho_1}{2\pi}\left(\frac{1}{r_1} + \frac{k}{r_2}\right) \tag{26}$$

If $\rho_2 < \rho_1$, the image is negative because k is negative and the poles 'attract', as shown on Fig.2.16. If $\rho_2 > \rho_1$, the image of A is positive and the current poles 'repel' one another, as seen on Fig.2.17.

If an observer is in medium 2, the previous boundary can be removed and the resistivity of medium 1 replaced with that of medium 2, but the strength of the source in the replaced medium must be altered (as was previously done when the observer was in medium 1) to $A(1-k)$, which is $A-kA$. Then, $A = I\rho_1/2\pi$, and the true source strength (that is, the current) is:

$$A = \frac{I\rho_2}{2\pi}(1-k) \tag{27}$$

as seen from medium 2. In medium 2, the potential at N, a distance r_3 from the source A is:

$$V_N = \frac{I\rho_2}{2\pi r_3}(1-k) \tag{28}$$

On the boundary between media, $V_1 = V_2$, and at a point r distant from both source and image:

$$\frac{I\rho_2}{2\pi r}(1-k) = \frac{I\rho_1}{2\pi r}(1+k)$$

and:

$$k = \frac{\rho_2 - \rho_1}{\rho_2 + \rho_1} \tag{29}$$

where k is also known as the resistivity image coefficient, or the 'reflection' coefficient. Here, 'reflection' may be a misleading term if it is taken in the geometric optical sense. The value of k varies between -1 and $+1$ over a complete range of resistivity contrasts.

The method of images is a useful way of determining potentials due to current sources near a resistivity boundary, and a family of type curves can be developed for solving field resistivity problems using this technique.

RESISTIVITIES OF EARTH MATERIALS

To obtain an acquaintanceship with any geophysical method requires a study of the physical properties of the substances being investigated. Other earth scientists rely strongly on optical characteristics at the other end of the electromagnetic spectrum for identifying earth materials. While electrical properties are measured in a macroscopic realm, they are usually due to very small scale phenomena.

Resistivity properties of minerals

The range of resistivities observed in geologic materials is considerable, being many, many orders of magnitude. Ideally, individual minerals are fairly consistent in their electrical characteristics, but their aggregate total range of resistivity is much greater than that of rocks.

Several true metals, that is, substances that conduct electricity by virtue of their high electron mobility, occur in native form. However, gold, copper, platinum, and silver are hardly common rock-forming minerals, even in the deposits where they occur. Unfortunately, graphite has metallic affinities that plague the prospector.

On an atomic scale, a metal can be considered an orderly packing of metal ions surrounded by an equal number of highly mobile valence electrons. The properties of a metal are due mainly to the large number of electrons permeating the material.

The energy necessary to start the electrons of a metal in motion is very low, especially if there are only a few imperfections in the crystal. The electrical resistivity of a crystal increases with the amount of packing disorder. Thermal motion of the metal ions also interferes with the motion of conduction electrons, and as temperature increases, the resistivity of true metals also increases.

A semiconductor is actually a nonmetal that also conducts by electron motion, but its resistivity is greater because there are fewer conduction electrons. However, the number of conduction electrons increases as temperature increases, giving a semiconductor a distinctive negative temperature coefficient of resistivity. Most of the common sulfide ore minerals are really semiconductors to some extent, but their resistivity may actually be as low as that of a true metal if their electrons have a low activation energy. These metallic-luster semiconducting minerals are sometimes called 'metallic minerals', but they really are not. Table IV (after Keller, 1966) lists some of the minerals with their conduction type and resistivity.

The form in which a mineral usually occurs, that is, its habit, is quite important in determining the manner in which the mineral affects the resistivity of a rock. Common rock-forming conductive minerals, such as magnetite, specular hematite, graphite, pyrite, and pyrrhotite, can lower resistivity considerably. Whether a mineral is platy or occurs in fractures filled with pore fluids will also have an important bearing on the resistivity of a rock. Sphalerite and hematite are both insulators, although iron-rich sphalerite (marmatite) and specular hematite are the more conductive.

Silicate minerals and other obviously nonmetallic minerals conduct electricity by ion charge transfer in the manner of an electrolyte. For this reason, they are called 'solid electrolytes', and as might be expected, they have very high resistivities. This conduction mechanism is not very important in most crustal rocks in mining areas because of the much better electrical conduction in nearby pore fluids, especially if these are in fractures.

Electrical conduction in rocks

The water content of most rocks almost completely determines the electrical conductivity, a fact expressed by Archie's law:

$$\rho_r/\rho_w = \phi^{-m} \tag{30}$$

where ρ_r is the resistivity of the rock containing pore fluid, ρ_w is the resistivity of the pore fluid, and ϕ is the rock porosity. The cementation factor m ranges from 1.4 for loosely cemented rocks to 2.5 for very tight materials and is taken to be 2.0 in Archie's law. Eq. 30 shows that the resistivity of rocks, especially intrusive types, is strongly dependent on porosity. The diagrammatic relationship between ρ_r and ρ_w is illustrated in Fig.2.18. The

TABLE IV

Resistivities of minerals (zero frequency)

		Reported resistivities (ohm-meters)	
		low	high
Metallic minerals			
Copper		$1.2 \cdot 10^{-8}$	$3.0 \cdot 10^{-7}$
Gold			$2.4 \cdot 10^{-8}$
Graphite	C, parallel to base plane	$2.8 \cdot 10^{-5}$	
Semiconducting minerals			
Argentite	Ag_2S	$8.0 \cdot 10^{-1}$	$2.0 \cdot 10^{-3}$
Arsenopyrite	FeAsS	$2.0 \cdot 10^{-5}$	15
Bismuthinite	Bi_2S_3	3.0	12
Bornite	$Fe_2S_3(3–5) \cdot Cu_2S$	$1.6 \cdot 10^{-6}$	$6.0 \cdot 10^{-3}$
Cassiterite	SnO_2	$4.5 \cdot 10^{-4}$	$1.0 \cdot 10^{+4}$
Chalcocite	Cu_2S	$8.0 \cdot 10^{-5}$	$1.0 \cdot 10^{-4}$
Chalcopyrite	$Fe_2S_3 \cdot Cu_2S$	$3.0 \cdot 10^{-5}$	$2.0 \cdot 10^{-1}$
Cobaltite	CoAsS	$1.0 \cdot 10^{-5}$	$1.3 \cdot 10^{-1}$
Covellite	CuS	$3.0 \cdot 10^{-7}$	$8.3 \cdot 10^{-5}$
Cuprite	Cu_2O	10	50
Enargite	Cu_3AsS_4	$2.0 \cdot 10^{-4}$	$9.0 \cdot 10^{-1}$
Galena	PbS	$6.8 \cdot 10^{-6}$	$5.8 \cdot 10^{-1}$
Hauerite	MnS_2	10	20
Ilmenite	$FeTiO_3$	$1.0 \cdot 10^{-3}$	4.0
Jamesonite	$Pb_4FeSb_6S_{14}$	$2.0 \cdot 10^{-2}$	$1.5 \cdot 10^{-1}$
Magnetite	Fe_3O_4	$9.3 \cdot 10^{-5}$	$1.0 \cdot 10^{+4}$
Manganite	$MnO \cdot OH$	$1.8 \cdot 10^{-2}$	$5.0 \cdot 10^{-1}$
Marcasite	FeS_2	$1.0 \cdot 10^{-3}$	$1.5 \cdot 10^{-1}$
Melaconite	CuO	$6.0 \cdot 10^{+3}$	
Metacinnabarite	4 (HgS)	$2.0 \cdot 10^{-6}$	$1.0 \cdot 10^{-3}$
Millerite	NiS	$2.0 \cdot 10^{-7}$	$4.0 \cdot 10^{-7}$
Molybdenite	MoS_2	$1.2 \cdot 10^{-2}$	$1.0 \cdot 10^{+6}$
Nicollite	NiAs	$1.1 \cdot 10^{-7}$	$2.0 \cdot 10^{-6}$
Pentlandite	$(Fe, Ni)_9S_8$	$1.0 \cdot 10^{-6}$	$1.1 \cdot 10^{-5}$
Psilomelane	$kMnO \cdot MnO, nH_2O$	$8.8 \cdot 10^{-2}$	$6.0 \cdot 10^{+3}$
Pyrite	FeS_2	$1.0 \cdot 10^{-5}$	$6.0 \cdot 10^{-1}$
Pyrolusite	MnO_2	$7.2 \cdot 10^{-3}$	31.8 ± 5.3
Pyrrhotite	Fe_7S_8	$2.0 \cdot 10^{-6}$	$1.6 \cdot 10^{-4}$
Rutile	TiO_2	29	$9.1 \cdot 10^{+2}$
Skutterudite	$CoAs_3$	$1.1 \cdot 10^{-6}$	$1.6 \cdot 10^{-4}$
Smaltite	$CoAs_2$	$1.1 \cdot 10^{-6}$	$1.2 \cdot 10^{-5}$
Sphalerite	ZnS	$1.8 \cdot 10^{-2}$	$4.0 \cdot 10^{+4}$
Tetrahedrite	Cu_3SbS_3	$3.0 \cdot 10^{-1}$	$3.0 \cdot 10^{+4}$
Uraninite	UO_2	1.5	$2.0 \cdot 10^{+2}$

Data from Keller (1966, table 26.1, p. 557—561).

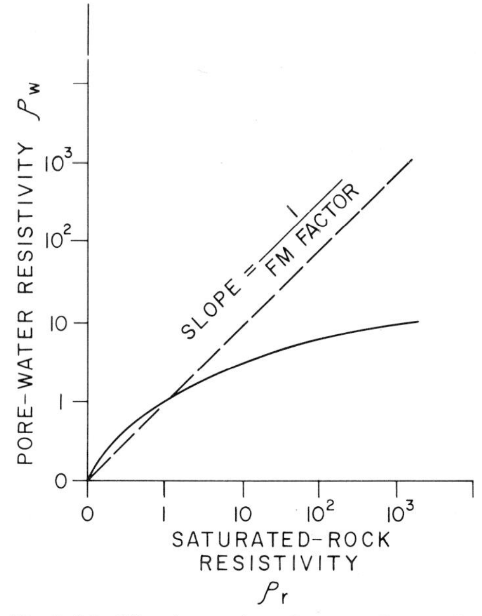

Fig.2.18. The formation factor of a rock plotted as a function of rock and pore-water resistivities.

ratio ρ_r/ρ_w is called the *formation factor* of the rock, which can be fairly distinctive for sedimentary rocks and therefore useful for correlation between drilled formations in petroleum exploration.

Since individual rock types vary considerably in their overlapping ranges of resistivity, this property is not a diagnostic feature. Figure 2.19 shows typical resistivity values of some general rock types. Geologic ages and structural history of a rock also affect its electrical characteristics.

The electrical conductivity of highly porous, water-saturated soils and semiconsolidated earth materials varies with the amount of pore fluid, as is implied by Archie's law. When the content of mobile ions in the pore fluid is high, as in arid regions, the resistivity of the alluvium and weathered rock is quite low. Where mobile ion content is low as in areas with better circulation of near-surface groundwater, resistivities are higher.

Resistivity of mineralized rocks

Except when an isotropic rock contains massive quantities (that is, above 10 or 15% by volume) of highly conductive minerals, the metallic-luster mineral content generally has only a minor influence on the rock's conductivity. For example, in an igneous rock with a fairly large amount of accessory magnetite, the magnetite grains tend to be disseminated, and since they are formed early in the rock's cooling history, there is seldom any adjoining electrolyte-filled pore space.

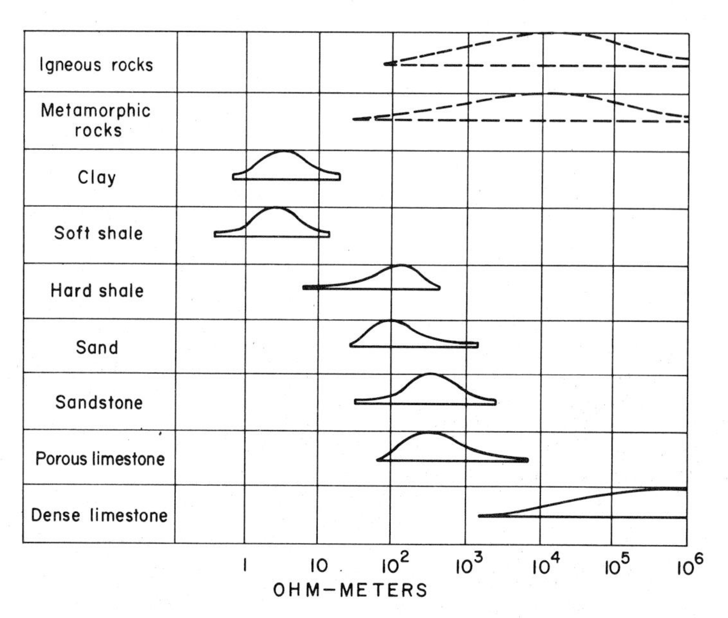

Fig.2.19. Range of typical resistivity values of general rock types.

The crystal habit of pyrite usually causes it to have a porous sugary texture, although large single crystals sometimes grow in vugs. Pyrite is a semiconductor, and it has an amazingly wide range of resistivity, averaging about 10^{-6} Ωm, which is similar to that of marcasite. Pyrrhotite, chalcopyrite, and arsenopyrite are good conductors with less variability in resistivity than pyrite. Their average resistivities are in the range 10^{-3}—10^{-4} Ωm. Magnetite and galena are very good conductors ($\sim 10^{-4}\Omega$m), but their isometric euhedral habits tend to isolate individual crystals, making the rock that contains them more resistive than might be supposed.

Graphite conducts electricity in two different ways, as a true metal (covalent bonding) along its basal plane and as an insulator (van der Waals bond) across it, with a direction variation in resistivity of four orders of magnitude from 10^{-6} to 10^{-2} Ωm. Also, pyrite is often found in a graphitic environment, which doesn't help the explorationists' search problem.

Keller and Frischknecht (1966) have observed that even some mineralized portions of American Smelting and Refining Company's Mission mine near Tucson, Arizona, have a fairly high resistivity. Figure 2.20 shows a mineralized section of a drill hole on this property and an electric log of the hole. Although this situation is unusual, it does point out that resistivity methods alone cannot be relied upon to identify the presence of economic mineral deposits.

Of course, some strongly foliated rocks, such as schists, have remarkable

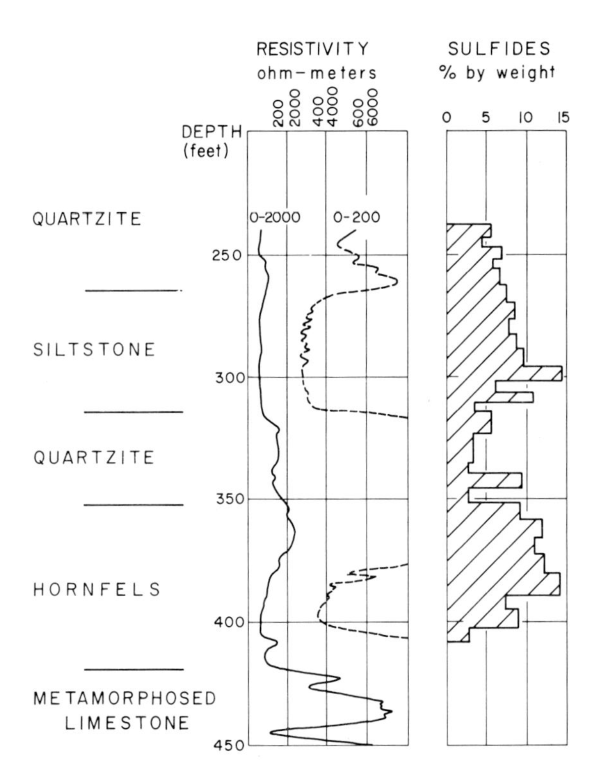

Fig.2.20. A mineralized section of drill hole of ASARCO's Mission mine near Tucson, Arizona, showing an unusual correlation of resistivity and percent mineralization.

continuous conductive veinlets, and these rocks, although probably anisotropic in behavior, can be expected to have a low resistivity.

It may be rightly concluded after seeing considerable field evidence of the causes of rock resistivities that electrolyte-filled pores and fractures in rocks tend to be much more continuous and extensive than are conductive mineral veinlets.

RESISTIVITY EXPLORATION METHODS

Field techniques in resistivity surveying have evolved to a considerable extent during the past half century. Large strides have been made in developing resistivity theory, and in so doing it has been necessary to gain an understanding of the electrical properties of rocks. Good instrumentation is a prime requisite for carrying out successful field surveys, and fortunately, progress in the development of equipment has been rather good.

Resistivity equipment

Modern resistivity surveying equipment consists of a power source, a transmitter (current sender), a receiver (voltmeter), and related peripheral apparatus, such as test meters, wire, electrode material, tools, etc. In some available field equipment, the transmitter and receiver are packaged together, and they may have mutually synchronous circuits. Nevertheless, it is necessary to view the transmitter and receiver circuits as being independent and coupled together only by the resistivity of the earth. The ratio of received voltage to impressed current times the geometric factor of the array and electrode interval is the measure of apparent resistivity. A schematic diagram for field resistivity equipment is shown on Fig.2.21.

The power source for the transmitter may be batteries whose voltage can be readily increased by an inverter and transformer. A gasoline motor-driven generator is probably more efficient than batteries in its power-to-weight ratio, if over 25 W power output is needed. Current from these power sources is usually stabilized, rectified back to dc (if it is not already dc), and chopped by a low-frequency switching circuit in the transmitter before being put in the ground. A fairly large transmitted current, which can be produced by a higher voltage or by making lower resistance contacts with the ground, is desirable.

The receiver is a battery-powered electronic voltmeter or potentiometer, which preferably does not need to draw current from the ground in order to make a satisfactory voltage measurement. Potential electrode contacts with the ground are often unglazed porcelain pots filled with a copper sulfate solution, which leaks through the pot to make electrical contact with the ground. A copper rod in contact with the solution does not develop a potential with its electrolyte salt in this nonpolarizing electrode.

A potential difference of up to one volt frequently exists in the field between the two porous-pot potential electrodes. Natural earth potentials, which are often larger than resistivity voltage signals to be measured, must be compensated for if the voltmeter is to be sensitive to transmitted potentials. This is done by inserting a variable voltage source in series with one input of the receiver, a self-potential (SP) buckout control circuit.

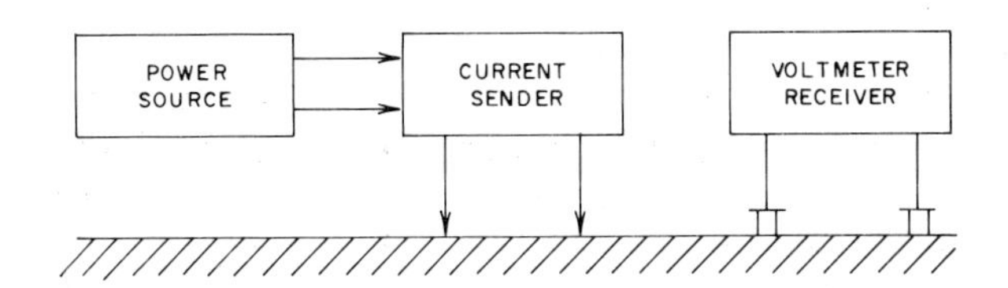

Fig.2.21. A generalized block diagram of field resistivity surveying equipment.

Self-potential

Natural potential differences between nonpolarizing porous-pot electrodes are mainly due to earth currents caused by regional telluric sources or are from subsurface electrochemical reactions that give rise to nearby self-potential (SP) or spontaneous polarization effects. After being measured, SP voltage variations can be interpreted and used as a geophysical exploration method. Although these SP voltages are meaningful, the method is by itself not a very highly developed prospecting technique because of a general lack of quantitative exploration significance for the anomalies. However, in areas where near-surface, massive sulfide bodies are being sought, the SP method gives the more direct results.

In earlier literature, SP currents were assumed to be due to a simple battery phenomenon involving oxidation of submetallic sulfides. Now it is apparent that two half-cell reactions are actually taking place: one above the water table and another below it. The metallic-luster sulfide body merely acts as an electrical conductor for the electrons flowing from the oxidizing to the reducing environment (Fig.2.22). SP anomalies due to the presence of subsurface sulfides are generally negative. Field survey values are plotted cumulatively, relative to one station or to some potential datum.

SP buckout control values from resistivity and IP surveys, even though sometimes of uncertain meaning, are worth plotting and interpreting. Considered with other exploration data, SP information is often helpful in the overall analysis of a subsurface situation.

Resistivity sounding

One practical application of resistivity surveying is found in regions with horizontally stratified materials in which resistivities are also stratified and relatable to the water table or other flat-lying geologic boundaries. Under these conditions, it is possible to make a resistivity sounding by making several apparent resistivity measurements using four earth-contact electrodes and varying the distance between electrodes. The process of making such a series of symmetric measurements about a point is called *resistivity sounding*, or making a *resistivity expander.*

It can be shown that for simple horizontally layered subsurface resistivities there is a direct mathematical solution to the surface-potential field pattern and that by evaluation of field survey data, the depth to resistivity interfaces can be determined and the true resistivities of the subsurface layers can be found. Determinations of depth and resistivity, if made directly from field data, are called direct interpretations. However, a set of type curves for a number of different subsurface situations is usually pre-computed or otherwise obtained and depths and resistivities found by com-

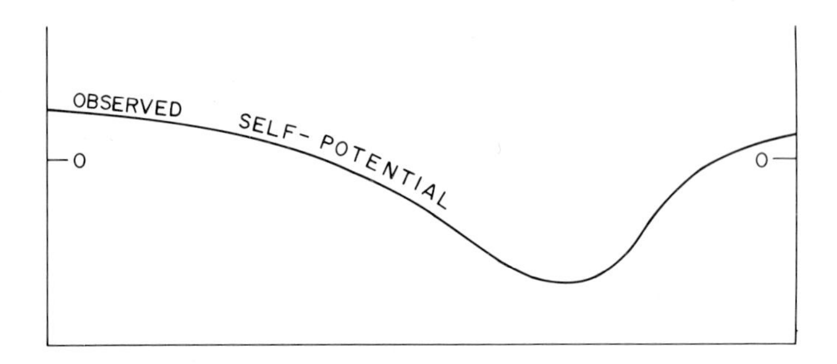

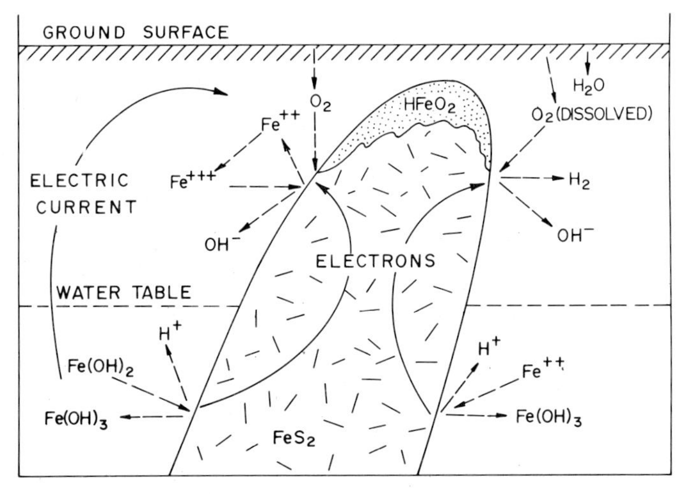

Fig.2.22. Self-potentials near an oxidizing massive sulfide body, showing current flow lines and chemical reactions.

paring field data with these catalogued curves. Figure 2.23 shows a calculated apparent resistivity diagram.

Field surveying using the resistivity sounding technique consists of setting out electrodes in a given configuration, initially with a very short interval between electrodes. Usually the resistivity obtained with this smallest electrode interval is assumed to be the true resistivity of the uppermost layer. The electrode interval is then enlarged or expanded successively about the center point by perhaps factors of two, and apparent resistivities are calculated for each of these spacings. As the electrode spacing is enlarged, the measured apparent resistivity approaches that of the lower layer(s). The plotted curve of apparent resistivity versus array interval, which is the apparent resistivity curve, is then compared with the family of precomputed apparent resistivity curves in order to obtain the most probable depth to an interface and the true resistivity of the materials below it.

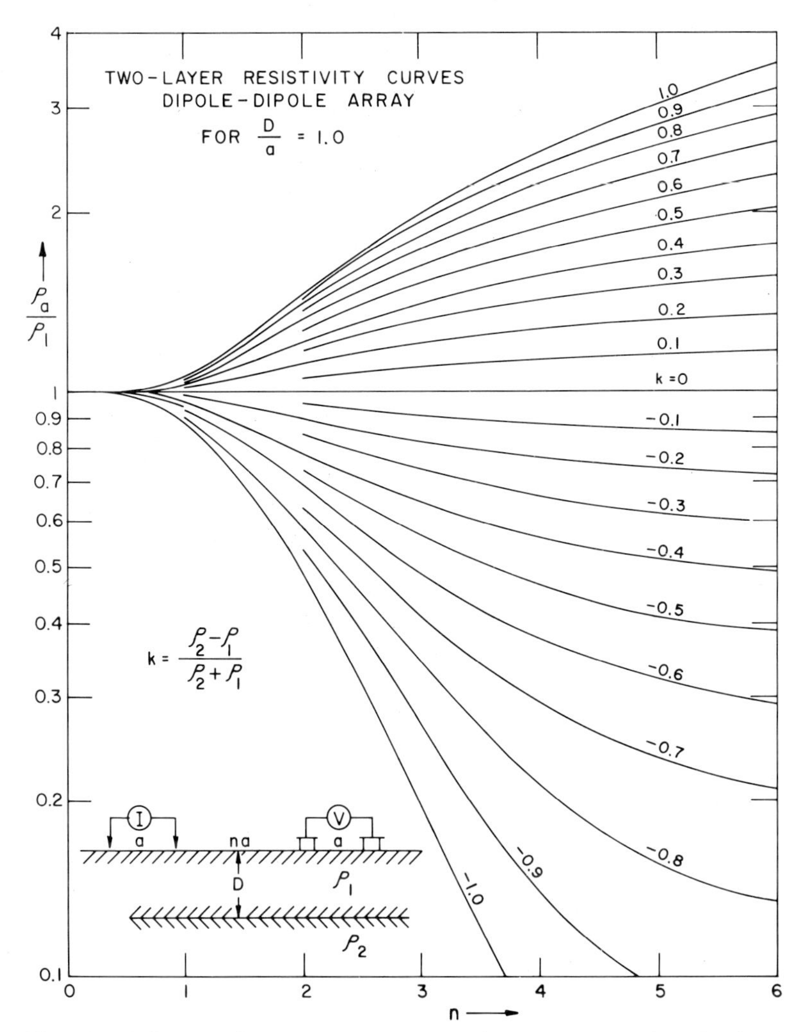

Fig.2.23. A family of two-layer apparent resistivity curves using the dipole—dipole array.

In practice, the apparent resistivity is usually divided by the small electrode-interval resistivity before plotting on the diagram, and the electrode interval is displayed on an arithmetic scale, as illustrated by the family of two-layer type curves shown on Fig.2.23. The shape of the apparent resistivity curve is dependent on the ratio of the resistivities. In a two-layer situation, there is a simple relationship between depth to interface and the curve inflection point for a fixed value of the resistivity contrast coefficient k. Ideally, the error in a depth measurement is the uncertainty in the exact location of the inflection of the field curve.

Three-layer and even four-layer situations can be interpreted by this curve-matching technique, although care must be taken to determine that horizontal layering and resistivity homogeneity within layers and near the surface actually exist before applying these interpretation techniques. Sets of resistivity-sounding curves have been compiled by La Compagnie Générale de Géophysique (1955), The Netherlands Rijkswaterstaat (1969), Orellano and Mooney (1966, 1972), and Lazreg (1972).

Resistivity profiling

In the practice of resistivity mapping in the field, soundings can be made at intervals and then the attempt made to correlate data between them. Another approach is to resort to a technique known as horizontal profiling. Here a fixed electrode interval is selected, perhaps from resistivity-sounding data, using an expander, and the array is traversed through successive survey points, giving a resistivity profile at the electrode interval. Sometimes field procedures can be worked out to run two or more electrode intervals on a single traverse by 'hitch-stepping' the array interval.

If vertical interfaces are sought in a resistivity survey, for instance, a fault or a dike, horizontal profiling is an effective search method. Certain sub-

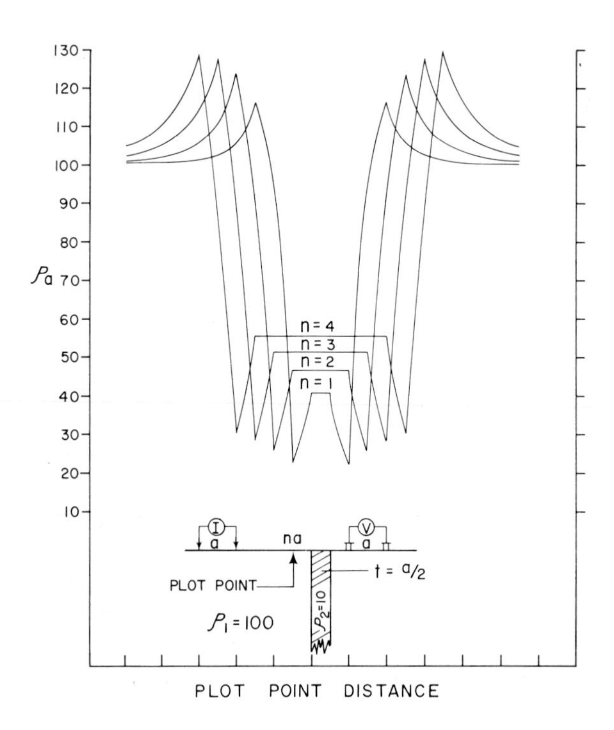

Fig.2.24. Horizontal profiles over a vertical dike using the dipole—dipole array.

surface shapes have distinctive resistivity profile patterns, like the one shown in Fig.2.24. If a near-vertical interface between contrasting resistivities comes to surface, the resistivity profile, shown here as a line connecting a series of very closely spaced measurements, will have a series of cusps, at the most one for each electrode as it crosses each vertical interface.

Pseudosection plotting

It is possible to plot both resistivity-sounding and profiling data on a single two-dimensional diagram, giving a pseudosection of apparent subsurface resistivity along a survey traverse. This is often called a resistivity pseudosection to emphasize the fact that these are only apparent values in terms of magnitude and location and that the diagram must be carefully interpreted before a true resistivity section can be constructed.

Resistivity values are plotted relative to conventions established for the array, at the midpoint of symmetric arrays and at a democratically established point on asymmetric arrays, and at a scaled depth related to the distance between transmitter and receiver. Those using the three-array prefer to use a point midway between the current electrode and near potential

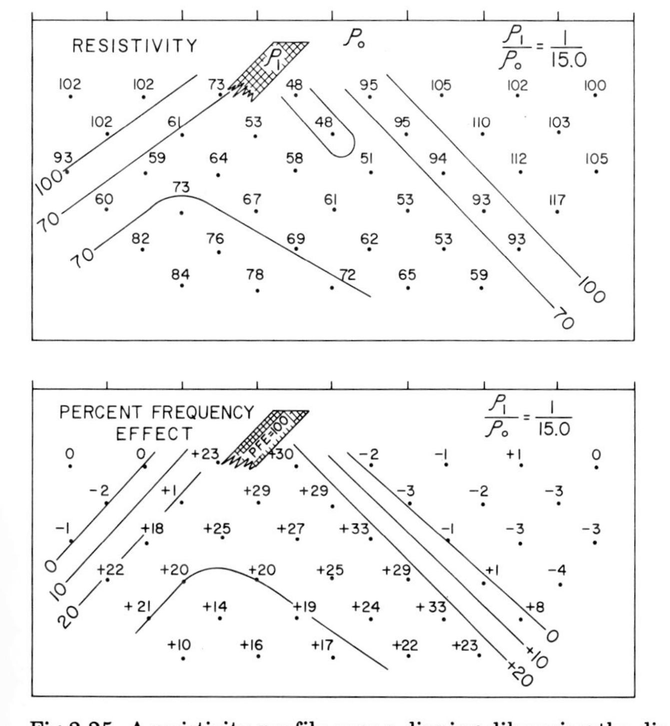

Fig.2.25. A resistivity profile over a dipping dike using the dipole—dipole array.

electrode as a plot point, and those using the pole—dipole array use a plot point that is halfway between the current electrode (pole) and a spot midway between potential electrodes. Plotted values are contoured, often at logarithmic intervals. Fig.2.25 shows the results of the plotting convention established for the dipole—dipole array. The pseudosection pattern does not necessarily give directly interpretable results. On Fig.2.25, the dike dips in a direction opposite to the trend of the resistivity low.

A resistivity traverse is usually laid out to intersect structural features at right angles, and a pseudosection can sometimes be crudely compared with an exploration cross section. Lateral resistivity variations are also sensed by the expanding array. This effect can confuse interpretation of the pseudosection.

Examples of resistivity sections over a dipping buried interface are shown in Fig.2.26. Note that the angle of dip cannot be readily interpreted except in a qualitative way at shallow dip angles.

Topographic features can give rise to fairly definite apparent-resistivity differences, even over material of uniform resistivity, especially using the dipole—dipole electrode array. A hill will be a resistivity high and a valley a resistivity low. Field data can be partly corrected by using diagrams of the sort shown as Fig.2.27.

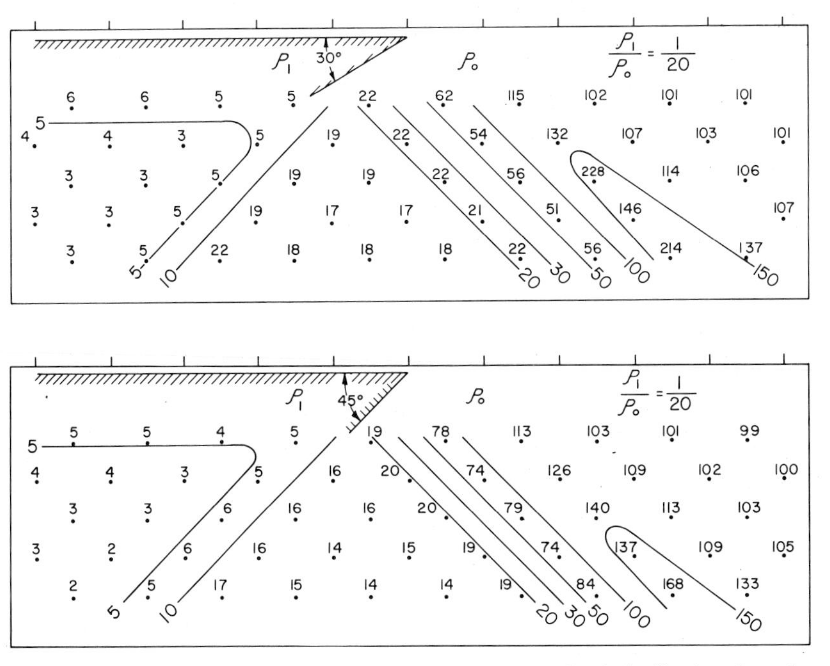

Fig.2.26. Plotted resistivity pseudosection over a buried dipping interface using the dipole—dipole array.

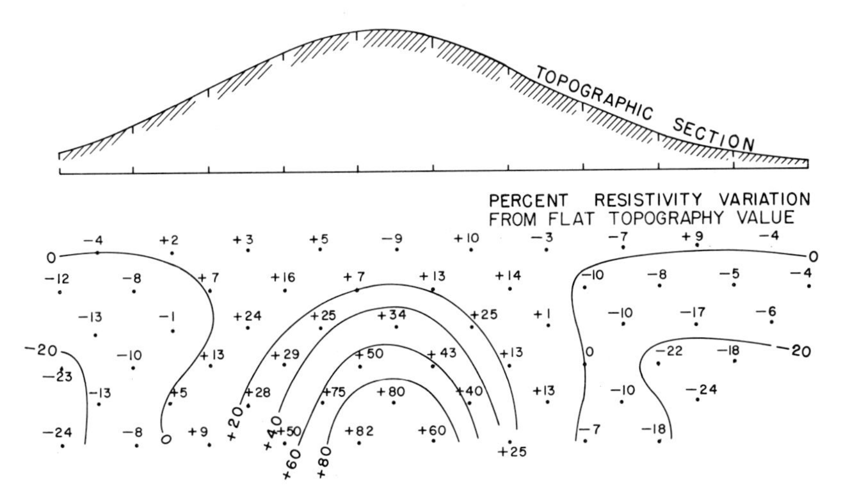

Fig.2.27. Plotted pseudosection over a topographic hill using the dipole—dipole array.

Chapter 3

THEORY OF INDUCED POLARIZATION

This chapter contains a review of the electrochemical conditions that exist between a solid and an electrolyte, especially those pertaining to the phenomenon of induced polarization. Also the mathematical theory of induced polarization will be presented.

ELECTROCHEMICAL THEORY

There are two basic classes of electrochemical phenomena that are concerned with natural liquid solutions and interfaces, such as are encountered in low-frequency electrical geophysical prospecting. Electrochemistry itself deals with the chemical changes and reactions produced by electricity, and electrokinetics is the study of the effects of electricity in motion. Although the electrochemical effect is by far the greater, both of these fields and especially their combined interactions are important in the understanding of the IP phenomenon.

The electrode

The reader should be aware that to an electrochemist an electrode is a contact that sustains a single, controlled electrochemical reaction, but to the geophysicist an electrode is a metal contact with the ground or a metallic mineral particle at which both oxidation and reduction may take place in a dirty system.

Electrochemists have long known that a potential difference exists across the interface between a metal electrode and an electrolyte, even when no net current flows across the interface. This potential drop at zero current flow is largely a reversible phenomenon, and though it cannot be observed directly it exists with virtually no energy loss in the system. Such an equilibrium potential is established by the ion concentration and exchange reaction next to the electrode itself. For a relatively insoluble metallic-luster mineral the net charge of ions going into solution is balanced by the net charge of ions coming out of solution. The ions going into and coming out of solution constitute a balanced electrical current flow, or 'exchange current'. The amount of exchange current, which is not easily measured, is proportional to the free energy of activation in the electrode reaction. The resulting reversible electrode potential is caused by the tendency of the metal to go into solution where unbound ions have more degrees of freedom.

It is not possible to measure the potential of a single electrode; a comparative measurement must be made with respect to a reference electrode. This creates observation and measurement problems. Ingenious electrochemical laboratory methods have provided a considerable amount of information about electrode phenomena. Fig.3.1 is a diagram of an interface between a metal electrode in equilibrium with a solution, potential being idealistically plotted against distance. The potential near an electrode is the sum of both chemical and spatial (electrostatic) potentials.

A fixed layer of ions is present in the solution next to the solid, although no current flows through the interface. These ions are firmly attached to the surface by electrical or chemical forces or both. The adsorbed relatively thin fixed layer creates an electrical capacitance when its electrical effects are taken together with near-surface electrons or ions within the solid, and this is a factor in electrical conduction at frequencies above dc. The mobile diffuse layer, which is next to the fixed layer, is initially electrostatically attracted to the fixed layer, and when current flows allows a surface conduction that lowers rock resistivities below those predicted by Archie's law (Fig.2.17).

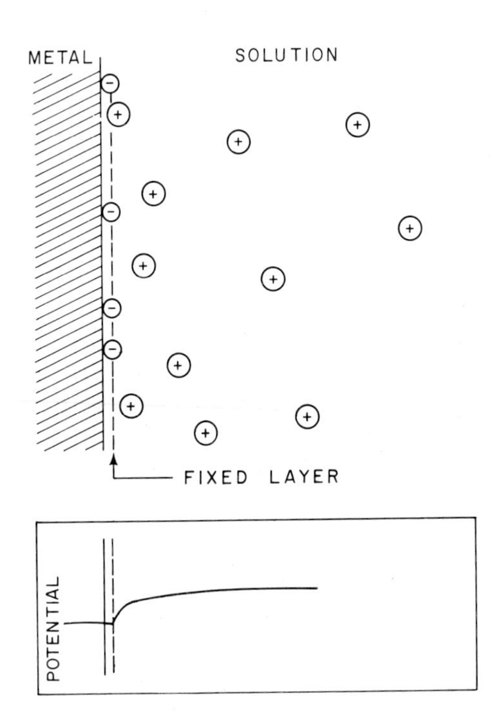

Fig.3.1. Reversible potential between an electrode and an electrolyte solution. Only ions in excess of electrical neutrality are shown. After Grahame (1947).

Overvoltage

When a small current is passed through to a metal-to-electrolyte interface the inherent potential difference changes. The difference in potentials across an electrode interface with a small current condition and the potential at a zero current condition, is known as the *overvoltage* or overpotential of the electrode and can be written in algebraic form as:

overvoltage = irreversible (observed) potential − reversible (unobservable) potential.

The overvoltage is the extra potential energy required to initiate an electrochemical process, particularly an electron-transfer reaction. It is mainly a potential due to an oxidation—reduction reaction termed 'activation overvoltage', and to a lesser extent a solution concentration gradient at the interface. Overvoltage is greatest at interfaces where the chemical activity is great, where the mode of conduction also changes from ionic in the electrolyte to electronic in the solid. At low currents, the overvoltage is observed to be proportional to the electric current density. The constant of proportionality is known as the 'polarization resistance' which is characteristic of the metal, the electrolyte, the temperature, and the direction of current flow. The overvoltage becomes increasingly nonlinear when the potential drop across the interface exceeds thermal energy levels, that is, above about $26/n$ mV. Here n is the number of molar equivalents.

The overvoltage at an interface may differ, depending on whether the current is going into or coming out of the metallic electrode. This is because there are different reactions involving oxidation at one interface and reduction at the other that take place at different rates. Thus, there is a net summation in overvoltage when one adds up the two reaction effects.

A voltage departure from the equilibrium potential at an electrode is also known as *electrode polarization*. This voltage departure in rocks, as in the laboratory, is due to interfacial electrochemical reactions and to ion accumulations in the pore-fluid near the surfaces of metallic-luster minerals. These ion accumulations oppose the current flow that creates them and are said to polarize the interface.

The double layer

Next to the electrode and the fixed layer there lies a layer of diffuse ions called the diffuse layer, whose existence and character are dependent on the flow of current through the interface. Ions in the diffuse layer may have the same or opposite sign as those of the fixed layer, depending on current density and the direction of current flow. Their number diminishes exponentially with distance from the interface, and the layer seldom extends past 100 Å from it. The presence of a double layer consisting of a fixed layer and a charged electrode will influence the overall conduction through the interface, causing a capacitance, which is effectively in series with the diffuse-layer capacitance. Like the electrode reaction impedance, the double-layer capacitance is frequency dependent at low frequencies. The capacitance due to the double layer ranges from 20 to perhaps as much as 300 $\mu F/cm^2$ of exposed electrode surface area. At higher frequencies, the double-layer impedance is nearly frequency independent. Using a simple model for the conduction phenomenon, the diffuse-layer ions can be assumed to move as point charges through a viscous medium under the influence of imposed electric fields or ion concentration gradients or both. Figure 3.2 is a display of calculated voltage variations next to an electrode. The potential of the electrode is linear with current density only when less than about $26/n$ mV of the electrode equilibrium potential.

The potential drop across the diffuse layer is known as the *zeta potential*, the effects of which cannot be directly observed. If there is an anomalous fluid pressure gradient through a porous material, such as a rock, fluid will flow through the porous medium. The energy of flow will be coupled to the electrical system inasmuch as ions are put in motion. An observable

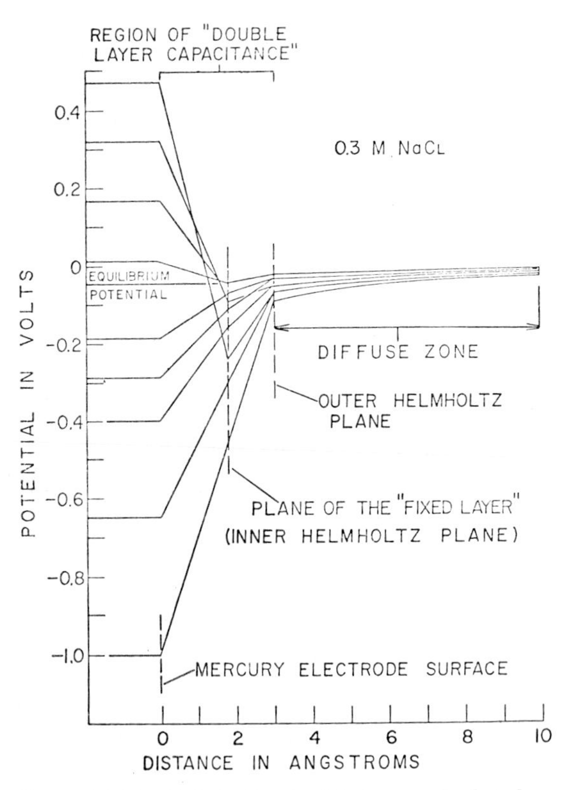

Fig. 3.2. Potentials in the double layer calculated at various current densities.

potential, called the streaming potential, can be generated by this electrokinetic means, and this sometimes contributes to the self-potential of rocks.

Seigel (1948) and Brant and Gilbert (1952) have pointed out that, according to electrostatic theory, overvoltage is directly proportional to current density for small current densities and reasonably short times. The simplified relationship they suggest is:

$$\eta = -\frac{4\pi\alpha}{K} sJt \tag{31}$$

where αJst is the strength of the double layer per unit area (α is the area of electrode surface, s is charge separation in the double layer, J is current density, and t is time, all in cgs units), and K is dielectric constant. Equation 31 does not take into account potentials due to the chemical reactions that are related to the double layer.

When an electrode has only a very small current passed through it, near-equilibrium conditions are maintained. If solution concentrations near the electrode are approximately unchanged overvoltage η is proportional to current density J by the relation:

$$\eta = -\left[\frac{RT}{nFJ_0}\right] J \tag{32}$$

where RT/nFJ_0 is called the 'polarization resistance', with units of ohm-cm^2 for an electrode one cm^2 in area. Polarization resistance is the inverse slope of the current—potential curve of Fig.3.3 at the point of zero current.

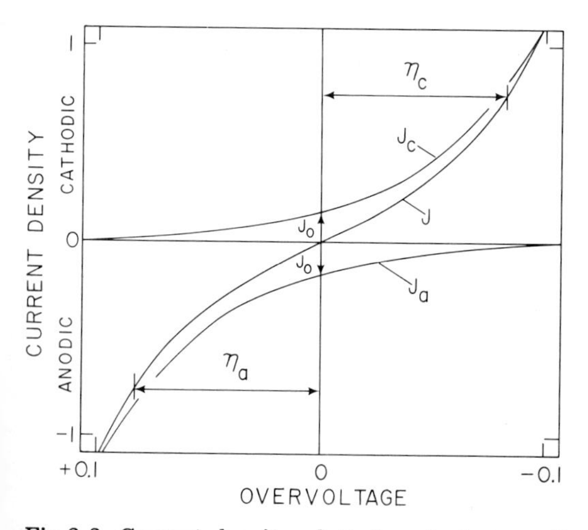

Fig.3.3. Current density plotted against overvoltage η. J is the algebraic sum of cathodic (J_c) and anodic (J_a) currents. Current density is in mA/cm^2, and potential in volts refers to the equilibrium potential. η_c and η_a are cathodic and anodic overvoltages. After Delahay (1954).

The factor RT/nF is expressed in volt-coulombs per equivalent faraday and has the value of 0.0257 volts per mole equivalent faraday at 25°C. J_0 is the exchange current density shown on Fig.3.3, a current which flows in both directions even though the net flow is zero. Equation 32 is related to the Boltzmann distribution law, which describes the tendency for thermal agitation to counteract the effects of electrical attraction and repulsion, and unlike eq. 31, mainly describes potentials due to oxidation—reduction reactions. The region of linearity of overvoltage with current density is limited to the central part of the curve of Fig.3.3. A larger current will produce temperatures at the interface above the ambient level and will cause nonlinear effects.

With longer charging times and larger current densities, Tafel's equation still holds:

$$\eta = a \pm b \log_{10} J \tag{33}$$

where a and b are thermodynamic factors with experimentally determined constants. The factors of eqs. 32 and 33 can be seen on Fig.3.3, with cathodic J_c and anodic J_a currents. The nonlinear relationship between η and J (but linear between n and the logarithm of current density J) holds above the IP current saturation threshold. Equation 33 applies even at low current densities, if the exchange current is taken into account. Figure 3.4 shows the distribution of unbalanced ions and potentials next to an electrode interface at larger current densities.

Although the overvoltage theory has been developed for the more reactive contact phenomena between a solid and an electrolyte, weaker surface potentials of a similar nature due to ion concentrations must also exist between ion-conducting solids and electrolytes, as will be discussed later.

Faradaic and nonfaradaic paths

Ideally, current can be carried by two different paths through an electrolyte in contact with an electron-conducting material. The first path is direct, by electrochemical charge-transfer involving oxidation and reduction, and diffusion of ions across the interface in what is called a faradaic electrochemical reaction path. The second path, consisting of charging and discharging the double layer in a nonfaradaic manner involving electronic capacitance, acts in a fashion similar to the displacement current in a conventional dielectric material.

There is an overall balance between the faradaic current, with its charge-diffusion and electron transfer reactions at the electrode, and the nonfaradaic current, which involves contributions from the capacitance of the fixed and diffuse layers. However, the rate of the nonfaradaic process, which is electrokinetic rather than electrochemical in nature, is much more rapid than the rate of the faradaic process. In nonfaradaic conduction, the diffuse

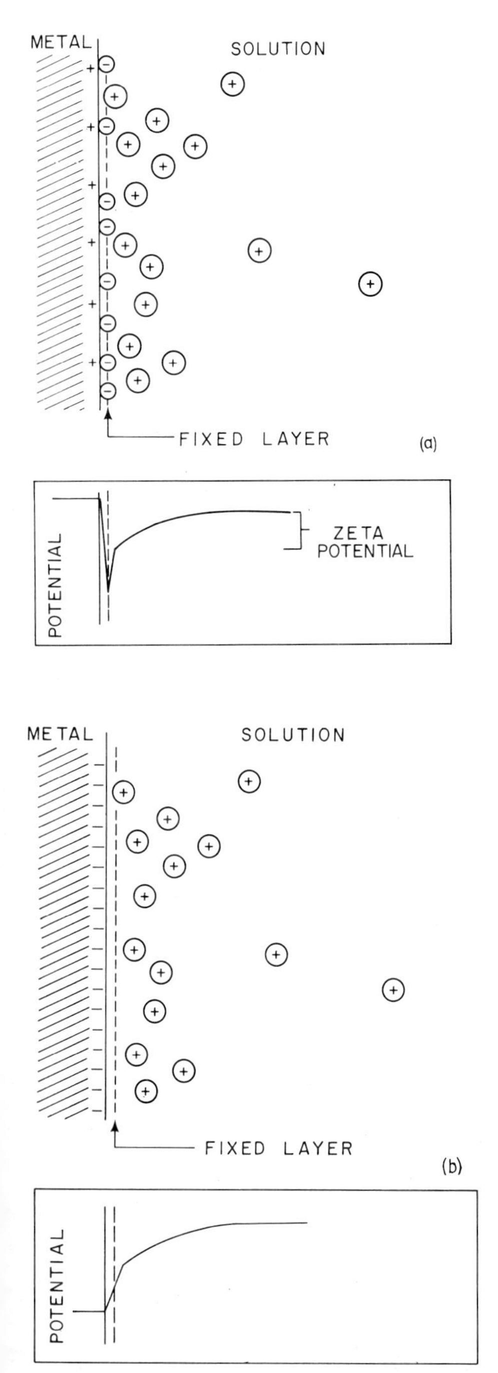

Fig. 3.4. Potential across the fixed and diffuse layers with (a) current into and (b) out of the interface, showing the region displaying what is known as the zeta potential.

layer is momentarily thinned by the electric field and the double layer behaves like a condenser in series with the current path through the solution. These paths are shown distributed diagrammatically by the parallel circuit in Fig.3.5, although a series circuit might do equally well. In rocks, the impedance of both of these paths is fairly linear with current density at very low current densities (less than 10^{-3} A/m^2) and at frequencies below 0.1 Hz. At higher frequencies, more current can be passed without appreciable nonlinearity.

Above frequencies usually used in field exploration, the nonfaradaic path carries much of the current, and its impedance or capacitive reactance is conventionally $J/2\pi fC$, varying inversely with frequency. At very high frequencies the nonfaradaic path may become frequency independent because the inertia of ions and their thermal velocity slow sympathetic oscillations and the resistivity of the material assumes a lower, relatively constant value.

Warburg impedance

The impedance due to the diffusion of ions in the faradaic path, which cannot be actually reproduced in a 'lumped' circuit with a simple finite combination of capacitors and resistors, is referred to as the Warburg impedance (Grahame, 1952). From the standpoint of circuit analysis, the Warburg impedance has the characteristics of a distributed electrical circuit. Its magnitude varies inversely with the square root of the frequency according to Grahame's derivation, $W = K(1 + J)/\sqrt{f}$. At low frequencies, it carries much of the current across the interface, and it becomes very large as f approaches zero. The circuit element symbol W is employed in an equivalent circuit network to point out the distinctive characteristics of the Warburg

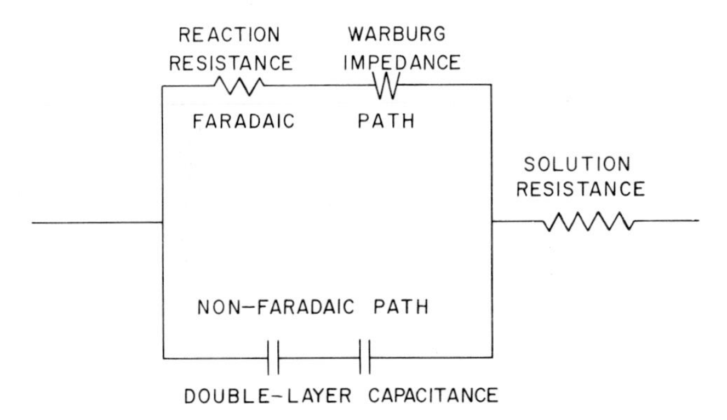

Fig.3.5. Faradaic and nonfaradaic current paths, showing these quantities as an equivalent electrical circuit across the interface between an electrolyte and electrode or electronic conductor.

impedance. W simplifies the more complex distributed aspects required to show electrode reaction impedance characteristics (Fig.3.6) of a mineralized rock.

Using the Warburg impedance relationship, it is possible to hypothesize the resistivity-against-frequency variation of an idealized circuit that behaves in a manner analogous to the IP characteristics of a rock. This resistivity spectrum (or conductivity spectrum) of the response behavior of a circuit is shown as Fig.3.7. The linear central portion of the plotted curve near its

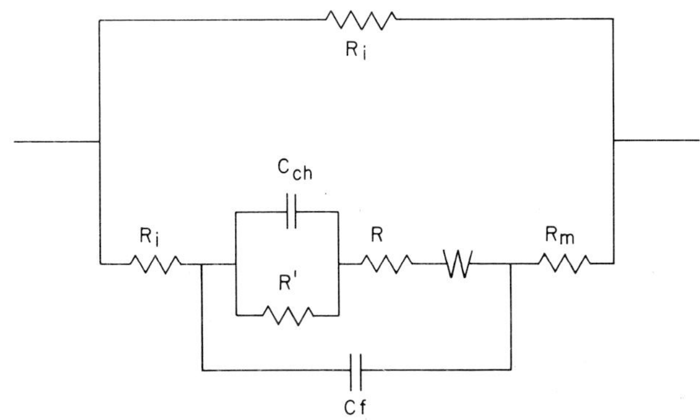

Fig.3.6. An equivalent circuit network showing a more complex circuit and the Warburg circuit element W. Here C_f is the double layer capacitance, C_{ch} is a chemical capacitance, R is reaction resistance, R' is resistance of higher order reactions, R_i is ionic path resistance, and R_m is the resistance path of the metallic vein or particle. After Ward and Fraser (1967).

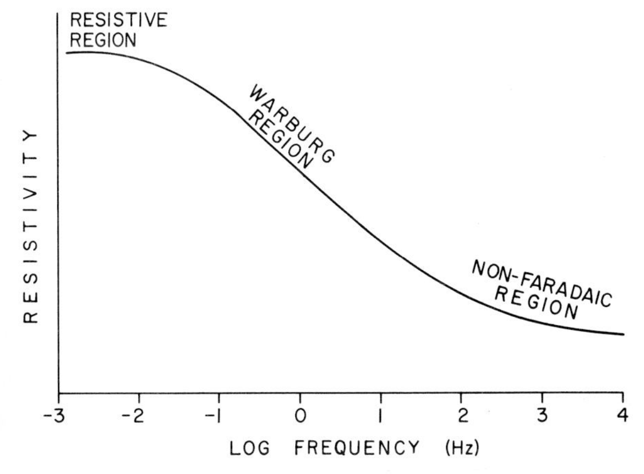

Fig.3.7. Idealized resistivity spectrum similar to the spectrum of a rock. In the central linear portion of the curve, known as the Warburg region, the slope of the curve over one decade of resistivity would be the IP effect, where resistivity can vary with an ideal maximum of $f^{-1/2}$.

inflection is known as the Warburg region. In this frequency band, the double layer is overridden, almost short-circuited, by the faradaic path. The slope of the spectral curve is the IP frequency response, which when normalized by dividing by resistivity is the frequency effect. A maximum frequency response can vary with resistivity as $f^{-1/2}$, but more often it is $f^{-1/4}$ because of parallel and distributed rather than only series polarization paths. Rock types with various kinds and amounts of mineralization (that is, disseminated, veinlets, or massive) will show somewhat different resistivity spectra, depending on whether the electrical properties of the rocks are characteristic of frequency measurements made above, within, or below the Warburg region (Fraser et al., 1964). Laboratory-derived examples of IP spectra will be reviewed in the next chapter.

Membrane polarization

Membrane polarization can be present, even though no current flows, when unsatisfied charges in clays or on cleavage faces or edges of layered and fibrous minerals attract a diffuse cloud of positive ions. These mobile ions can block pore passages and will migrate and accumulate in the presence of an electric field. The capillary pores of a rock also appear to sort electrolyte ions according to size and charge, some rocks behaving like an ion diffusion membrane.

Membrane polarization is superficially like electrode polarization, but no chemical reaction takes place at the solid-to-solution interface. It is not caused by a faradaic process but is due to ion accumulations within the electrolytic conductor. It is found under conditions in which there is a variation in ion mobility causing an ion 'pile up' at boundaries between regions where ion mobility varies.

When driven by an electric field, the randomly distributed positive ions in the electrolyte can readily pass through the positive ion cloud of the diffuse layer that is characteristically present in the clay-electrolyte system, but the drifting negative charges accumulate in this ion-selective membrane. Figure 3.8 shows ions in a pore forming a zone of ion concentration and creating membrane polarization. Constrictions in the pore space can also sieve out ions according to size and reduce the number by crowding, as at a turnpike entryway.

Membrane polarization is another means of storing electrical energy, by the difference of mobilities of ions. Where pore fluid is confined in a rock, excesses or deficiencies of ion populations will occur at boundaries of zones with different ion mobilities. The ion concentration gradients thus developed oppose current flow and cause a polarizing effect. Warburg impedance, existing here to some extent as in electrode polarization, represents a frequency-sensitive diffusion impedance resulting from the presence of membrane zones.

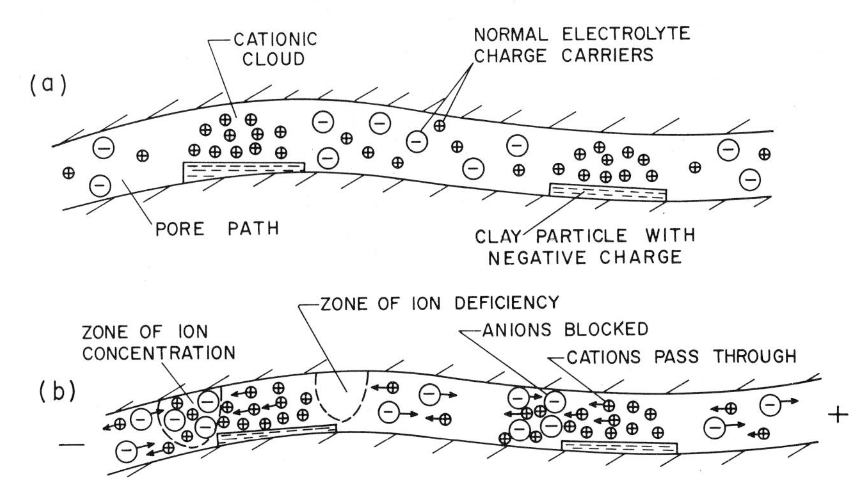

Fig.3.8. Depiction of ions in a pore space forming an ion concentration barrier which creates membrane polarization. (a) Pore path before application of an electric potential. (b) Pore path after application of a direct-current driving force. After Ward and Fraser (1967).

Most of the current in an electrolyte is carried by the positive diffuse-layer ions, and if the pore passages in a rock or mineral are physically constricted there will be a low mobility for negative ions. This situation is often found in the clay minerals that have very small passageways between sheet structures. Therefore, these ion-selective membrane zones of clays must occasionally be alternated with more open pore spaces in order for polarization to take place because there is so little movement of negative ions in the intramineral pore passageways. The length of the large pore passageway should also be small if the time scale of polarization is to be in a readily observable range. Dirty sands and a few rock types containing fibrous and layered minerals exhibit these required conditions and also give rise to membrane-polarization effects. These minerals have an abundance of small pore passages and a large exposed surface area. The surface area of many of the naturally occurring clay minerals is reported to be as much as several hundred square meters per gram (Grim, 1968). Table V is a list of some rocks and minerals that appear to have caused an above-background IP effect. The principal measurable difference between membrane and electrode polarization is the magnitude of response. Membrane IP effects are much smaller. An exception to this is the possible membrane-polarization effect that occurs near metallic minerals (Madden and Cantwell, 1967) where mobile ion species unrelated to the metallic mineral will effectively block current flow, giving a much larger polarization than ordinarily expected from usual membrane-polarization materials.

TABLE V

Nonmetallic rocks and minerals reported to give an above-background IP effect

Rock or mineral	Location	Investigator
Dirty sands	New Mexico, California	Vacquier et al. (1957) Bodmer et al. (1968)
Crystal lithic tuff	Safford, Arizona	Madden and Marshall (1958)
Tremolitic limestone	Willcox, Arizona	Madden and Marshall (1959a) Keller (1959)
Blue limestone	western Nevada	Heinrichs (1967); Van Voorhis and McDougall (1959)
Whitetail conglomerate	San Manuel, Arizona	anonymous
Serpentinite and asbestos	Canada	Seigel (1967)
Cummingtonite schists	Homestake mine	Sauck (1969)
Montmorillonite	Yugoslavia	Sumi (1961)
Zeolite beds	Bowie, Arizona	anonymous
Basalt flows	Port Arthur, Ontario	anonymous
Clay	Jerome, Arizona	Mayper (1959)
Basalt with fibrous minerals in vesicles	Safford, Arizona	Laboratory of Geophysics, University of Arizona
Titaniferous magnetite	Kelso, California	Elliot and Guilbert (1974)

APPLICATIONS OF ELECTROCHEMICAL THEORY

The electrochemical concepts of reaction behavior and ion-concentration gradients near a surface provide a factual one-dimensional model for the IP mechanism. Inevitably, natural physical and chemical systems are more complex, but the fundamentals described in the preceding part of this chapter can be directly applied to a better understanding of observed IP phenomena.

With a background knowledge of the phenomena, one can have an appreciation of basic IP processes and behavior. Polarization is understood to be created by an excess of accumulating ions focused at interfaces, and this mechanism reduces the amount of current flow. Electrochemical reactions and charge diffusion largely influence the magnitude of polarization.

When current flow ceases, Coulomb's-law forces between the ions that have piled up in the polarization process cause the ions to start returning to their equilibrium positions. This relaxation constitutes a charge flow that can be measured as a decaying polarization potential. Certainly, the exact electrochemical reactions and ion diffusion and mobility conditions are not yet well known in the natural system.

Dielectric constant

Details on the dielectric property of materials are even less well known or understood than their conductive behavior. A dielectric substance is one that exhibits a polarization due to charge separation in an electric field. If the material is electrically homogeneous, isotropic, and linear, this phenomenon is expressed as:

$$P_d = K_0 \chi_d E \tag{34}$$

where P_d is charge dipole moment per unit volume in coulomb-meters per meter^3, K_0 is the permittivity (capacitivity) of free space, χ_d is electric susceptibility, and E is the electric field intensity in volts per meter. Ideally, the dielectric polarization P_d is proportional to the applied electric field. The constant K_0 is equal to $8.85 \cdot 10^{-12}$ F/m, and χ_d is a dimensionless dielectric property of dielectric material. Conventional dielectrics have a very high resistivity and are used as electric insulators.

The dielectric constant K of a material is the ratio of the capacitance C of the capacitor with the material between its plates to the capacitance C_0 of a similar capacitor in a vacuum:

$$K = \frac{C}{C_0} = \frac{q/v}{q_0/v_0} \tag{35}$$

The charge on the plates and the voltage between them are q and v with the material between the plates and q_0 and v_0 in a vacuum between the plates. The larger the dielectric constant of the material, the larger the charge that will be present on the capacitor.

The increased charge on the capacitor plates is present because of the charge polarization that occurs within the dielectric material. There are a number of ways in which this can occur, but they all result in the relative displacement of positive and negative charges within the material. Different types of dielectric polarization usually take different lengths of time to form, so that each can be separately observed below a characteristic frequency. The general categories of dielectric polarization are illustrated in Fig.3.9.

Interfacial polarization illustrated in Fig.3.9d can occur in electrically heterogeneous materials with contrasting dielectric constants and conductivities, where charge builds up near interfaces. This concept is similar to that of induced polarization except that it does not take into account the chemical polarization reactions or the conductive nature of natural earth materials.

Electric fields are propagated in insulating materials at the speed of light accompanied by what are known as displacement currents. This mode of conduction by displacement currents is minimized in electrically conductive solids because induced countercurrents cause transient electric fields to be readily attenuated with distance. Energy is rapidly expended in transmitting

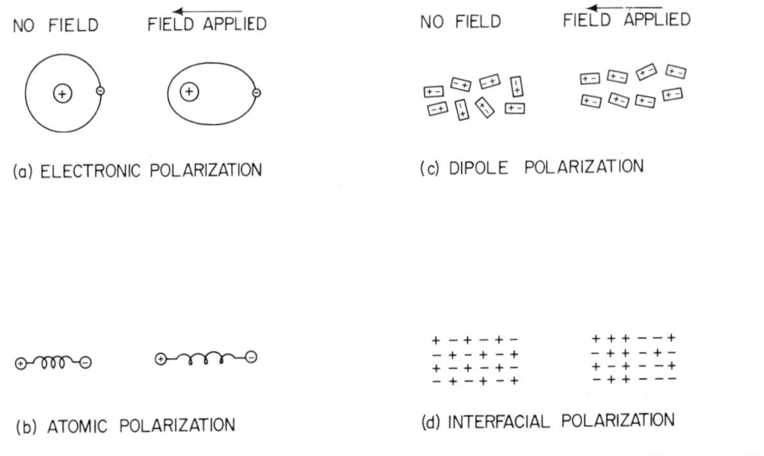

Fig.3.9. Different types of dielectric polarization. After Von Hippel (1954).

high-frequency electrical energy through conductive substances. At high frequencies, the dielectric 'constant' becomes an index of refraction and dispersion occurs.

Except for metallic minerals, dry rocks are good insulators and are also dielectrics, sometimes exhibiting the interfacial polarization of Fig.3.9d. However, in nature, rocks are very seldom dry enough to be considered true dielectric materials. Measurement of dielectric constant and electrical conductivity of moist rock at frequencies from 10^2 to 10^6 Hz has been made by Scott et al. (1967). By dielectric and insulating standards, rocks are fairly good conductors, and from this can be inferred that rock in place is wet.

ELECTRICAL MODELS OF INDUCED POLARIZATION

A logical three-dimensional model of a polarizable rock with disseminated mineralization has been put forth by Seigel (1949) and Wait (1959), who consider polarized, spherelike, metallic particles as dipoles. At a constant low current density and for a given volume of metallic mineral, the induced current dipole moment per unit volume P is inversely proportional to particle size. Also, the decay constant is related to particle size for a simple electric and geometric model of a polarizable rock.

Although the IP phenomena are basically electrochemical in nature, the systems are really too complex to be represented by a single set of chemical and thermodynamic relationships. The types of ions, their concentrations, and the reactions involved are only generally known. The pore geometry and its relationship with metallic mineral particles are also important. It is quite difficult to represent all these factors explicitly. Thus, it is convenient to use simplified macroscopic analogies in the way of either 'lumped' or 'distributed' electrical circuits.

From an electrical standpoint, IP phenomena are at the very low frequency end of the electromagnetic energy spectrum, and magnetic inductive effects are negligibly small, as are currents due to electric displacement. The inductive electromagnetic coupling effect is an exception to this statement, and it will be treated in a later chapter.

For the most part, direct-current theory and Laplace's equation will hold, and application of alternating-current theory, except where necessary to clarify the physical measurement problem, need not be invoked.

Equivalent circuit

A simple electric circuit consisting of an arrangement of resistors and capacitors will not show a response identical to that of the IP effect, but in many ways the IP behavior of mineralized rocks can be approximated with the lumped equivalent circuit shown in Fig.3.10. An analysis of such a circuit helps in gaining an insight into the more complex electrical features of rocks. However, even if the circuit performance is identical to field observations, there is no assurance that a phenomenon is uniquely explained. Similar circuit analyses apply to most simple electrical devices.

The elementary parallel equivalent circuit of Fig.3.10 has a purely resistive arm R_{dc}, representing the behavior of resistive ionic conduction in unmineralized current paths near a metallic mineral. The resistor R_b can be compared to the resistance due to blocked conduction paths within mineralized rock, and the capacitor C can be associated with the double-layer capacitance and Warburg impedance.

An additional resistor in series with the parallel circuit of Fig.3.10 would be analogous to an unpolarized conduction path and would provide for an instantaneous drop in potential from V_p to V_s.

At low frequencies, the capacitor is an open circuit and a resistance measurement shows only R_{dc}. At higher frequencies, the capacitor approaches a short circuit and a resistance determination depends on both R_b and R_{dc}. Thus, the change in circuit resistance (impedance) is a measure

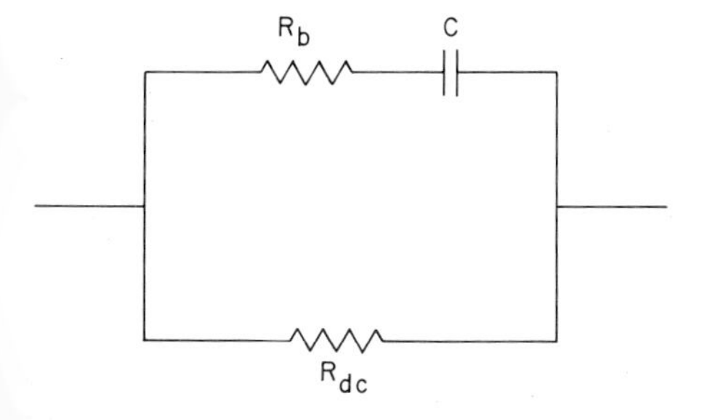

Fig.3.10. Equivalent circuit of the electrical current paths through and near an interface.

of the effect of blocked conduction paths as the frequency f of the current is varied.

The impedance Z of the parallel circuit is:

$$\frac{1}{Z_f} = \frac{1}{Z_{dc}} + \frac{1}{Z_b}$$

and:

$$Z_f = \frac{Z_b Z_{dc}}{Z_b + Z_{dc}}$$

Since:

$$Z_b = R_b + \frac{1}{j\omega C}$$

then:

$$Z_f = \frac{R_{dc}\left(R_b + \dfrac{1}{j\omega C}\right)}{R_{dc} + R_b + \dfrac{1}{j\omega C}} \tag{36}$$

where $\omega = 2\pi f$.

If the frequency effect FE of the circuit is defined as the percent impedance difference between two frequencies, Z_f minimum and Z_f maximum (Fig.3.11), then:

$$FE = \frac{Z_f \text{ minimum} - Z_f \text{ maximum}}{Z_f \text{ maximum}} \tag{37}$$

One can substitute impedance values in eq. 37, using Z_f minimum = R_{dc} and Z_f maximum = Z_f of eq. 36, and the solution is:

$$FE = R_{dc}/R_b \tag{38}$$

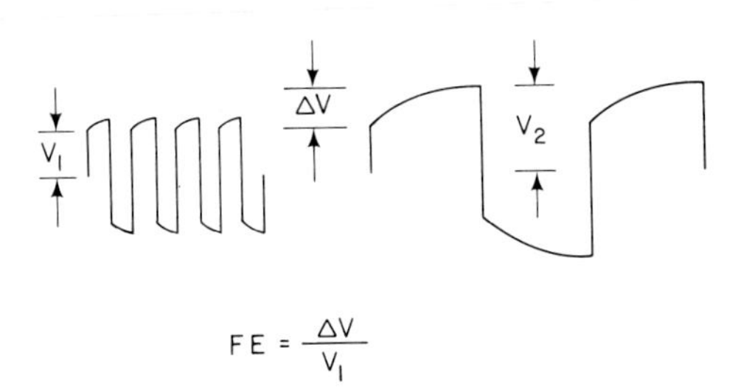

Fig.3.11. Voltages at low and high frequencies, shown on an unfiltered frequency type of waveform. The inducing current at each frequency is constant.

The FE becomes larger with more numerous blocked parallel conduction paths and R_b becomes smaller. It can be seen that in the frequency method the capacitor is essentially a switch. The switch is actuated by the frequency of the current source. The value of the capacitance C could be determined by using a spectrum of frequencies but not by using just two frequency values.

A time-domain analysis detects the decaying voltage (Fig.3.12) across R_{dc} resulting from the discharging of capacitor through resistances R_b and R_{dc}. The transient voltage observed is expressed as:

$$V_t = V_p \left[\frac{R_{dc}}{R_{dc} + R_b} \right] e^{-t/\tau} \tag{39}$$

where $V_p = IR_{dc}$ and the time constant $\tau = (R_b + R_{dc})C$. This analysis suggests that the more resistive but equally polarizable materials should display longer time constants. The secondary voltage V_s is measured at time $t = 0$ and is normalized by the primary voltage V_p, defining chargeability M as the ratio of V_s/V_p:

$$M = \frac{V_s}{V_p} = \frac{R_{dc}}{R_{dc} + R_b} \tag{40}$$

At small values of FE and M, these fundamental IP parameters are essentially equal, that is:

$$FE \approx \frac{M}{1 + M} \tag{41}$$

If one measured the integrated area under a decay voltage curve normalized by V_p, this quantity is diagnostic of the value of the capacitance C because the time constant of the decay voltage is given by $(R_{dc} + R_b)C$. A larger capacitor leads to a longer decay time and also a larger integrated

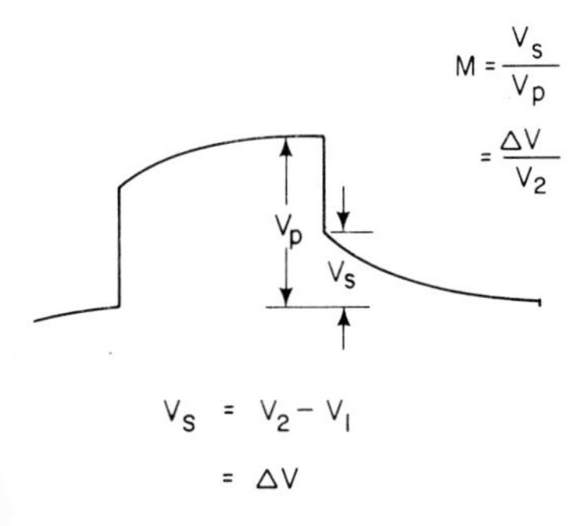

Fig.3.12. Primary (V_p) and secondary (V_s) voltages on an IP waveform. Equations show an equivalence of time and frequency IP measurements.

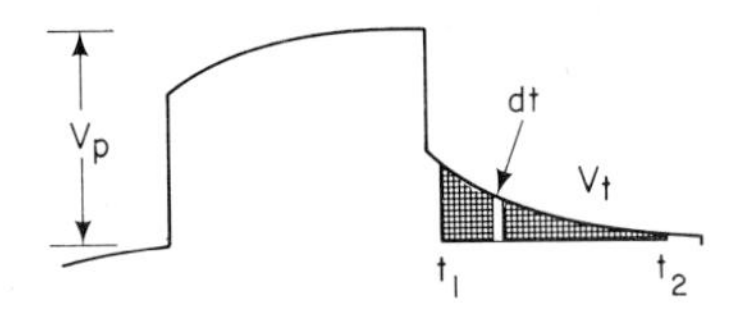

Fig.3.13. The integrated decay voltage that is also used as a measure of chargeability M.

signal. This integrated voltage has also been called the chargeability M (Fig.3.13):

$$M = \frac{1}{V_p} \int_{t_1}^{t_2} V_t \, dt \qquad (42)$$

and from our equivalent circuit $M = R_{dc} C$. Generally, the time constant for mineralized rocks ranges between 0.1 and 10 sec, with most values between 0.2 and 0.6 sec. Note that in the foregoing study, measurement of the integrated signal depends directly on the capacitance C. If C is a diagnostic parameter of polarizable rocks, it could be estimated by dividing M by R_{dc}, giving a quantity closely related to the metal factor or specific capacity.

Analysis of this simple circuit consisting of lumped elements thus shows three values, FE, M, and C, for the impedance properties of materials that can be measured. The significance of the decay constant and metal factor are noteworthy. There is sometimes a difference of opinion about which parameter is the most meaningful. Although a circuit analysis tells quite a bit about electrical behavior, it cannot answer the question of the relative importance of these measurements values in rocks beyond showing an equivalence between time- and frequency-domain methods.

Other types of equivalent circuits can be devised to show the mechanisms of the faradaic and nonfaradaic paths and membrane polarization (Madden and Marshall, 1958; Ward and Fraser, 1967), and they are useful in displaying in simple terms the operation of complex phenomena. Indeed, the model of the Warburg impedance is based on a circuit of this sort. The circuit of Fig.3.10 could be appropriately modified by replacing the capacitance C with the Warburg circuit element W in which the impedance varies as $f^{-1/2}$ or $f^{-1/4}$, giving a much better fit to the decay curve. Equivalent circuits can also be formulated in which the electrical elements present in rocks are distributed over the wide range of values. Electrochemically, a mineralized rock behaves as if it were made of a semiporous electrode material. A 'ladder network', which is a type of distributed element circuit designed by Dolan and McLaughlin (1967), helped them in designing field equipment. A circuit analogy by Zonge (1972) shows that it is possible to approximate closely the frequency dependence of the Warburg impedance by a relatively simple distributed circuit element (Fig.3.14). Using reasonable circuit values, one can study the exponential or logarithmic frequency dependence and decay waveform factors.

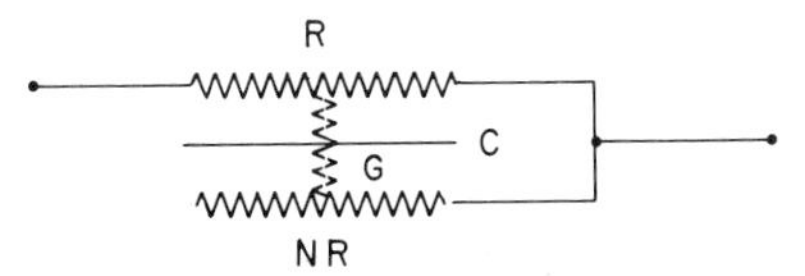

Fig.3.14. A distributed circuit diagram demonstrating an IP-like phenomenon. R is the dc resistance, C is capacitance, N is the ratio of the résistance through the ac paths to the dc resistance, and G is the leakage conductivity of the equivalent capacitance. $C \propto 1/f^{\frac{1}{2}}$.

The decay curve

The form of the IP decay curve is characteristic of electrical and mechanical relaxation phenomena. This fact gives some indication of the nature of the IP mechanism.

As suggested by the analysis of the lumped circuit, the basic IP decay curve can be shown as an exponentially decreasing shape of the form:

$$V_t = V_s \mathrm{e}^{-ut} \tag{43}$$

where V_t is the decay voltage at any time t after the secondary voltage V_s, and u is the reciprocal time constant of the curve $u = 1/\tau$.

It has been suggested by several who have studied IP data that an IP decay curve actually consists of several components, each with different time constants, given, for example, by Keller (1959) as:

$$V = V_s \sum_{n=1}^{n} A_n \mathrm{e}^{-u_n t} \tag{44}$$

where A_n is the amplitude of the nth term and u_n is the reciprocal time constant for the nth term. This equation can be expressed as an integral:

$$V_t = V_s \int_0^{\infty} G(u)\, \mathrm{e}^{-ut}\, \mathrm{d}u \tag{45}$$

where $G(u)$ is a distribution function defining the density of time constants in the interval u to $u + \mathrm{d}u$. The integral of eq. 45 can be evaluated for a lognormal distribution of time constant values, which is a type of function frequently observed in natural phenomena. To carry out this distribution analysis, V_t/V_s can be tabulated and treated as an inverse Laplace transform using an appropriate approximating function.

It is possible to show the IP decay curve to be composed of a Fourier series of trigonometric functions. This process allows a close fit of the mathematical expression with observed IP data. Also, the shape of the IP decay curve can be recorded in digital form, which can then be computer processed with a time-to-frequency transform function and displayed as a resistivity spectrum.

Wait (1958) and Scott and West (1969) have noted that the primary IP decay curve has a shape which is logarithmic. On the early part of the curve the decay voltage drops almost immediately from V_p to V_s (Fig.3.12) and then decreases logarithmically with time, whereas the exponential decay of a simple equivalent circuit (Fig.3.10) would be continuous from V_p with no distinctive change in slope at V_s.

Departures of the shape of the IP decay curve from its general logarithmic form may hold implications about the nature of the mineralization of a sample or of the subsurface. For example, according to whether the mineralization is disseminated, veined, or massive, the distribution function curves may have a characteristic symmetry or skewness that is also suggested from study of resistivity spectra. However, just because a curve can be exactly fitted to observed field data does not necessarily explain the meaning of the data. A word of caution should also be given concerning the possible distorting presence of aberrant decay curves in field data due to the interaction of decay currents from adjacent polarizable bodies and from the negative IP effects in certain geometric situations.

Field of a dipole

In dielectric materials that are insulators, an analysis of the field of two unlike charges separated by a force of an electric field is a necessary first step in studying dielectric properties. An argument may be advanced that the IP effect does not actually induce a uniform distribution of electric dipoles in a rock. Also, there are those who may not wish to classify IP phenomena with dielectric behavior of materials. Nevertheless, it is sometimes convenient and useful to draw on the complete analogy between a conducting dielectric and a polarizable substance. As mathematical models, the comparison is exact in that we are studying the volumetric phenomena of induced dipole moments. This approach is similar to the method of images discussed in Chapter 2 inasmuch as static charges in insulating materials can be replaced by current sources and sinks in a conductive medium. Of course, these current sources and sinks are virtual rather than real.

The dipole shown in Fig.3.15 is in a medium of resistivity ρ and consists of positive and negative induced current sources, $+I_i$ and $-I_i$, of equal magnitude separated by a distance s. This distance is small compared with the distance r to the point O at which we wish to find the electric potential V.

At O:

$$V = \frac{\rho I_i}{4\pi}\left(\frac{1}{r_b} - \frac{1}{r_a}\right) \tag{46}$$

where:

$$r_a^{\,2} = r^2 + \left(\frac{s}{2}\right)^2 + sr\cos\theta$$

and:

$$\frac{r}{r_a} = \left[1 + \left(\frac{r}{2r}\right)^2 + \frac{s}{r}\cos\theta\right]^{-1/2}$$

Expanding:

$$\frac{r}{r_a} = 1 - 1/2\left(\frac{s^2}{4r^2} + \frac{s}{r}\cos\theta\right) + \frac{3}{8}\left(\frac{s^2}{4r^2} + \frac{s}{r}\cos\theta\right)^2$$

or, if we neglect terms of order higher than s^2/r^2:

$$\frac{r}{r_a} = 1 - \frac{s}{2r}\cos\theta + \frac{s^2}{4r^2}\left(\frac{3\cos\theta^2 - 1}{2}\right)$$

Similarly:

$$\frac{r}{r_b} = 1 + \frac{s}{2r}\cos\theta + \frac{s^2}{4r^2}\left(\frac{3\cos\theta^2 - 1}{2}\right)$$

hence:

$$V = \frac{\rho I_i s}{4\pi r^2}\cos\theta \quad (r^2 \gg s^2) \tag{47}$$

Note that the potential due to a dipole falls off as the *square* of the distance r, whereas the potential from a single current source (eq. 17) varies only as $1/r$. This comes from the fact that the induced current sources in the dipole appear close together to an observer some distance away and their fields cancel more and more as the distance r is increased.

A dipolar-type current field is approximated in the dipolar current electrode configuration when the observer is a considerable distance from the current electrode pair.

A uniformly polarized sphere gives the exact field of a dipole, and with a

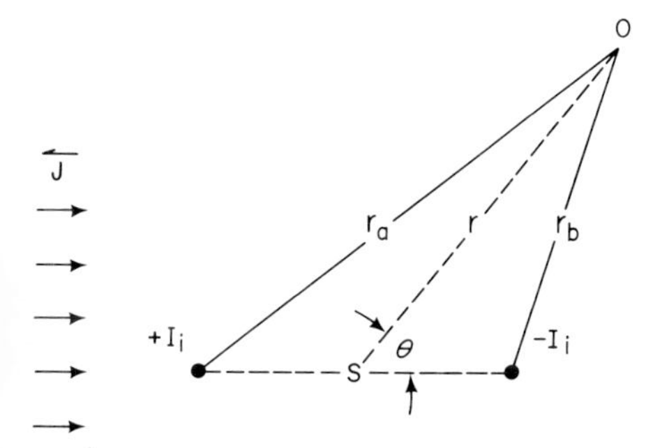

Fig.3.15. A current dipole in a resistive medium. Although the inducing-current field $\overleftarrow{J}$ is shown, it is not included in eq. 46 for the development of the potential of the current dipole.

summation of a volumetric distribution of small polarized spheres, any polarization property of materials can be approximated. Of course, a highly conducting sphere cannot be truly internally polarized, but surface polarization effects can be simulated using the relationships discussed next in this chapter.

ELECTRICAL POLARIZATION

The product of the induced current source I_i and the distance between the sources is the current dipole moment $\overleftarrow{p} = I_i \overleftarrow{s}$, which is a vector quantity (Fig.3.16), whose magnitude is $I_i s$ and which is directed from negative to positive current source. The number n of dipole moments $\overleftarrow{p}$ per unit volume of a material defines the induced current polarization moment P of the material:

$$\overleftarrow{P} = n\overleftarrow{p} \tag{48}$$

where P is the average induced current dipole moment per unit volume in ampere-meters per meter^3. Ideally, the time and space over which P is defined must be large enough such that there is not a great fluctuation in P from one time interval to the next or from one volume to a neighboring one. P is analogous to the 'intensity of magnetization' of magnetic materials.

Polarization

Next to be considered is the change in electric potential at a point brought about by nearby polarization. The properties of a polarized solid can be viewed both internally and externally. The current sources and sinks induced in and on the polarizable body are virtual rather than real. The exterior inducing current field is $\overleftarrow{J}$, whose potential is not included in the following development.

The interior of the body is taken here to be uniformly polarized as shown

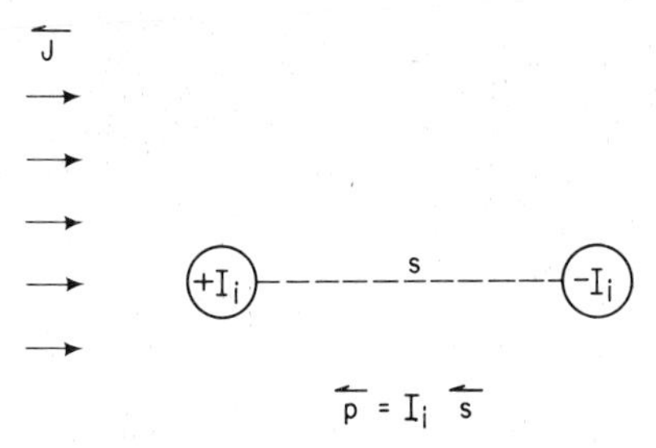

Fig.3.16. An induced-current dipole, illustrating dipole moment. The inducing-current field is $\overleftarrow{J}$. The current dipole moment $\overleftarrow{P}$ is a vector quantity directed from $-I$ to $+I$.

in Fig.3.17. Exterior to the body, the change in potential resulting from uniform total polarization of the body is the same as if only the existing surface charges or virtual current sources remained. One might anticipate this phenomenon by noting that interior dipolar charges or current poles are all internally satisfied and during the depolarization process the internal currents need not be seen at the surface.

The equivalence between volume polarization replaced by surface poles on a body is demonstrated mathematically by vector calculus. From eq. 47, the three-dimensional potential dV at a point O due to an elemental volume dv of dipoles is:

$$dV = \frac{\rho}{4\pi} \overleftarrow{P} \cdot \nabla \left(\frac{1}{r}\right) dv \tag{49}$$

Integrating the potential at point O due to a total volume of dipoles:

$$V = \frac{1}{4\pi} \int_v \rho \overleftarrow{P} \cdot \nabla \left(\frac{1}{r}\right) dv \tag{50}$$

where $P\,dv$ is the dipole current strength of a volume element and ∇ is the gradient operator taken at the volume element.

Using the identity:

$$\nabla \cdot \left(\frac{\overleftarrow{P}\rho}{r}\right) = \frac{1}{r} \nabla \cdot (\overleftarrow{P}\rho) + \rho \overleftarrow{P} \cdot \nabla \left(\frac{1}{r}\right)$$

then:

$$V = \frac{1}{4\pi} \int_v \nabla \cdot \left(\overleftarrow{P}\rho \frac{1}{r}\right) - \frac{1}{r} \nabla \cdot (\overleftarrow{P}\rho)\, dv \tag{51}$$

And by Gauss's theorem, the first integral term is replaced by the surface integral:

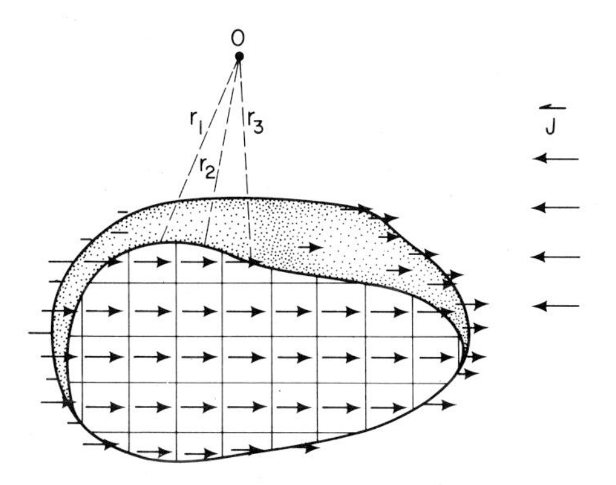

Fig.3.17. Section through a uniformly polarized body with elemental volume dipole moments. Inducing-current field is $\overleftarrow{J}$.

$$V = \frac{1}{4\pi}\int_s \frac{P_\perp \rho}{r}\,ds - \frac{1}{4\pi}\int_v \frac{1}{r}\,\nabla\cdot P\rho\,dv \qquad (52)$$

where s is the surface bounding the volume v.

Thus a volume distribution of current dipoles is mathematically equivalent to a surface distribution of strength P, the normal component of the polarization P out of the surface s (Fig.3.18), plus a volume distribution of current sources of density equal to $-\nabla\cdot P$. The second term of eq. 52 vanishes if the current sources and sinks within the body are balanced, which usually occurs in induced polarization. The change in external polarization potential is due only to the virtual current sources and sinks on the surface, as indicated in the first term.

Equation 52 provides a useful boundary condition for treatment of polarizable bodies that is analogous to the application of eq. 22 for resistive bodies. The difference between normal components of current density $J_{\perp in}$ and $J_{\perp out}$ at an interface with a polarizable body can be equated at the surface. This is illustrated in Fig.3.19:

$$J_{\perp in} - J_{\perp out} \propto P_\perp$$

Another consequence of eq. 52 is the conclusion that the net sum of the total surface charges on a polarized body is usually zero. That is to say, when the total quantity of surface current sources is equal to the quantity of surface current sinks, their net scalar sum is zero.

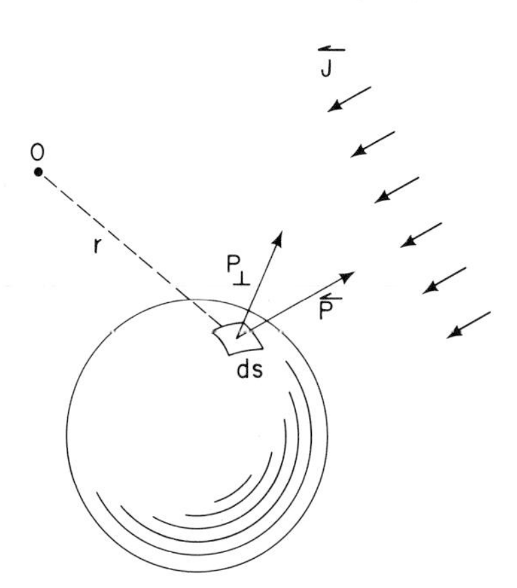

Fig.3.18. A polarized sphere, showing vectors $P_\perp$ and $\overleftarrow{P}$ at the surface element ds. The exterior inducing-current field is $\overleftarrow{J}$.

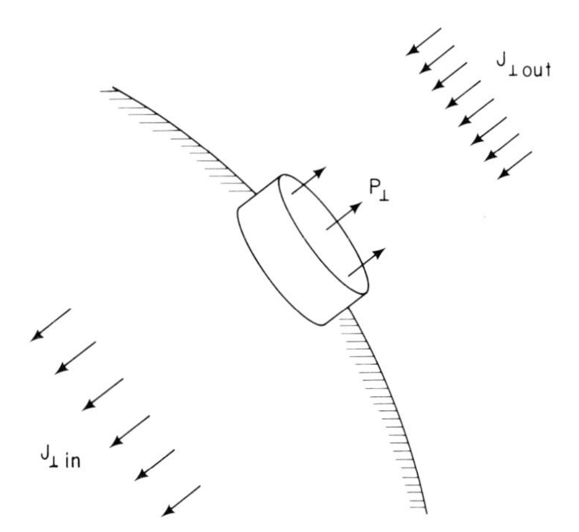

Fig. 3.19. Perpendicular current flow J in and out of an interface through a Gaussian surface.

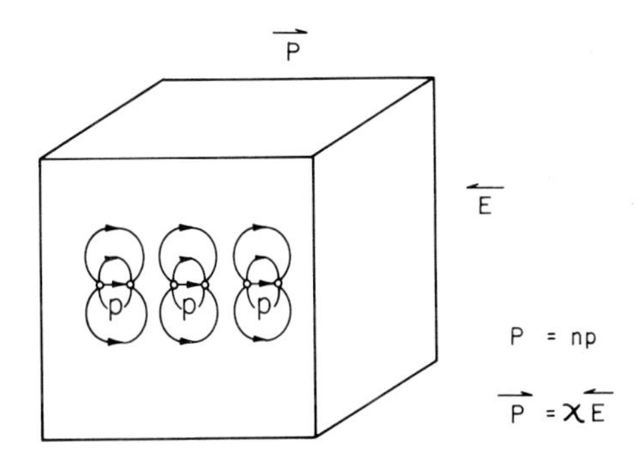

Fig. 3.20. A conducting dielectric (polarizable body), showing the external field $\overleftarrow{E}$ and polarization $\overleftarrow{P}$.

Induced polarization

If a conducting dielectric is in an external electric field E, it will become polarized by an amount P, as shown in Fig. 3.20. The vector quantities E and P will be directly proportional if the material is treated within its ohmic and dielectric range of behavior, but E and P act in opposition because the steady-state induced current dipoles are antiparallel to the inducing field E. Therefore:

$$\overleftarrow{P} = -\chi \overleftarrow{E} \tag{53}$$

where E is in volts per meter and χ is a constant of proportionality whose dimension is mhos per meter. This formula is the basic description of a conducting dielectric material, polarization $\overleftarrow{P}$ being proportional to the stimulating electric field $\overleftarrow{E}$. The constant χ is the electric susceptibility of the polarizing material, analogous to the magnetic susceptibility of magnetic materials.

In evaluating the meaning of the constant χ we must call on other relationships involving P and E. It has been pointed out that the IP phenomenon is essentially linear with current density at low current densities as far as electrochemical theory is concerned. Laboratory measurements by Collett (1959) and Henkel and Van Nostrand (1957) show nearly linear variation of induced polarization with current density, whereas McEuen et al. (1959) and Anderson and Keller (1964) find some variations. Seigel (1959a) has formulated a relationship based on the proportionality of polarization with current density, and he has defined chargeability M in terms of induced dipole moment per unit volume P as:

$$\overleftarrow{P} = -M\overleftarrow{J} \tag{54}$$

where $\overleftarrow{J}$ is current density in amperes per meter^2 and chargeability M is a dimensionless constant that is a basic polarization measurement characteristic of a homogeneous isotropic material. The minus sign signifies that the induced current dipole moment $\overleftarrow{P}$ shown on Fig.3.21 is in the opposite sense to the current density vector $\overleftarrow{J}$. Equation 54, the fundamental law of induced polarization, is a steady-state relationship wherein polarization $\overleftarrow{P}$ is proportional to the inducing current density $\overleftarrow{J}$. Chargeability (the constant of proportionality) M is a basic physical constant of the polarizable material related to polarization resistance of eq. 32. Overvoltage is seen to act in a fashion similar to polarization.

From eq. 54, chargeability $M = -P/J$, and $\overleftarrow{P}$ is proportional to but in an opposite sense from the secondary voltage V_s, with $\overleftarrow{J}$ proportional to the primary voltage V_p, so that theoretically, a fundamental observable measurement of the IP effect is:

$$M = V_s/V_p \tag{55}$$

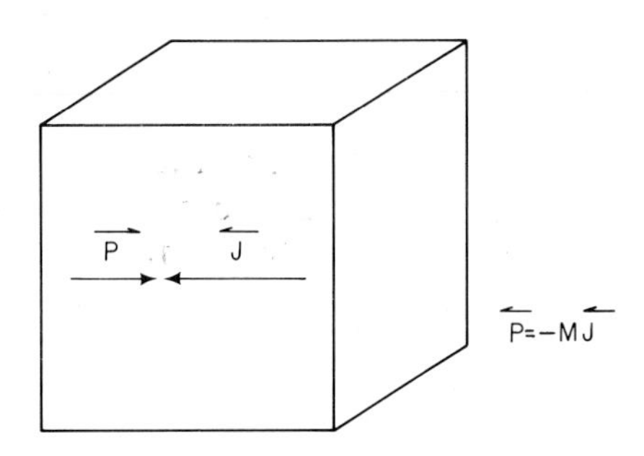

Fig.3.21. Polarization $\overrightarrow{P}$ and its spatial vector relationship with current density $\overleftarrow{J}$. The length of the vector $\overleftarrow{J}$ is proportional to $\overrightarrow{P}$ by the amount of chargeability M.

In actual field practice, chargeability is often measured as an area under the decay curve rather than by the defining equations 54 and 55. Using a pulsed square wave with a three-second cycle (+, 0,—, 0, with a 12-second period), a one-second integration time from the start of the decay voltage will give a chargeability (M_{331}) equivalent to the theoretical M. Most pulse field measurements are calibrated to this standard value. Usually, the field measurement of chargeability is given in units of millivolt seconds per volt, or milliseconds, which multiplies the M given in eqs. 54 and 55 by 1,000.

Metal factor

Ohm's law, written to include the dimensional properties of a material, was given in Chapter 2, as eq. 13:

$$\overleftarrow{E} = \rho \overleftarrow{J}$$

Combining this with eq. 54, gives:

$$\overleftarrow{P} = -\frac{M}{\rho} \overleftarrow{E} \quad (56)$$

which on comparison with eq. 53 proves that:

$$\chi = M/\rho \quad (57)$$

The quantity χ is also the *metal factor* of the frequency domain or *specific capacity* of the time domain. Metal factor (MF) is therefore a polarization quantity relating dipole moment per unit volume $\overleftarrow{P}$ and inducing field strength $\overleftarrow{E}$. When measured in mhos per meter, the ratio, M/ρ in mks units, is multiplied by $2 \cdot 10^3$ in order to convert to the English units of the metal factor (MF), as originally defined by Madden (1957). He gave the metal factor as:

$$\text{MF} = \frac{\rho_{dc} - \rho_{ac}}{\rho_{dc}\rho_{ac}} \cdot 2\pi \cdot 10^5$$

where resistivity ρ is measured in ohm-feet. The numerical multiplying factor is used to put the metal factor into the same numerical range as resistivity and chargeability. Some groups have called the metal factor 'metallic conduction factor', abbreviated as MCF, to emphasize the 'conduction change' nature of the metal factor.

Electric susceptibility (or specific capacity) χ is related to the dielectric constant K by normalizing χ to relative susceptibility χ_0 so that it is dimensionless, whereby:

$$K = 1 + \chi_0$$

By substitution:

$$M/\rho = K - 1 \quad (58)$$

which is to say that specific capacity, or metal factor, is related to the dielectric constant due to induced polarization.

IP phase angle

For purposes of illustration, it is sometimes convenient to represent polarization as a momentarily 'frozen' impedance vector in the phasor convention of the electrical engineer. A phase or 'phasor' diagram can be used to display polarization as the departure from the zero-frequency reference vector. This diagram can also be laid out in terms of the equivalent circuit of Fig.3.10. The reference vector direction is the 'real' or 'in-phase' axis, and the imaginary or 'out-of-phase' direction is at quadrature in a direction of 90 degrees. At zero frequency, the in-phase polarization component is taken as zero or unity, but the physical model (Fig.3.21) of the IP phenomenon shows that there is truly a steady-state, in-phase polarization. This difference is really only in the frame of reference. The out-of-phase (imaginary) polarization component at zero frequency is assumed to be zero, and an equivalent circuit analogy indicates that as frequency departs from zero to small phase angles (less than 10 degrees) the in-phase and out-of-phase components are comparably small.

In displaying IP phase data in graphical form, it is convenient to make a plot, such as suggested by Cole and Cole (1941), on which in-phase is shown against out-of-phase response (Fig.3.22). The resultant curve is formed by phase voltage ratios at successive frequencies and the shape of the curve is distinctive of the polarization behavior of the subsurface.

The frequency IP method measures the difference between impedance components at two or more frequencies without regard to phase in accordance with eq. 37. A phasor diagram is illustrated as Fig.3.23a. The polarization phase angle β is negative because the IP voltage lags the inducing current.

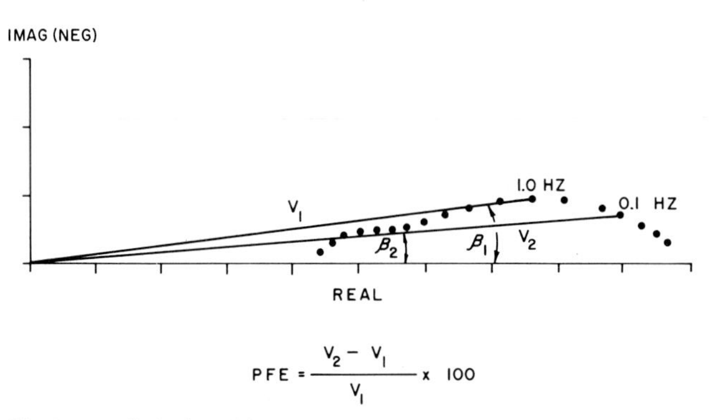

Fig.3.22. Relationship between frequency-domain induced polarization and complex resistivity.

In the pulse IP technique, the secondary voltage V_s and the transient voltage V_t are determined during the current-off cycle; therefore, at the frequencies normally used for IP work, the IP response phenomenon can be looked at as essentially an out-of-phase or quadrature measurement, especially when P is much smaller than J. The quadrature polarization vector P_Q would then be shown on a rotating vector diagram as leading the resistive component J_R of the total inducing current density vector J by $\pi/2$ or 90 degrees, as shown in Fig.3.23b. Then from eq. 54:

$$\tan \beta = P_Q / J_R$$

and:

$$\tan \beta = M$$

Also, when the tangent of the angle is less than 10 degrees, $\overleftarrow{P} < \overleftarrow{J}$, and the tangent and sine are approximately equal to the angle itself, thus:

$$\beta \approx -M \tag{59}$$

In normal polarization conditions using a one-cycle (1 Hz) frequency, a phase shift of 5.5 milliradians is equivalent to a change of one-percent frequency effect measured over a one-decade frequency interval ($\log_{10} f_H/f_L = 1$). These polarization units are about the same as a chargeability M of 5 millivolt seconds per volt measured with the standard pulse IP technique of three seconds on, three seconds off, and using an effective integration time of one second, written as M_{331}.

An IP phase-angle spectral plot when compared with a resistivity

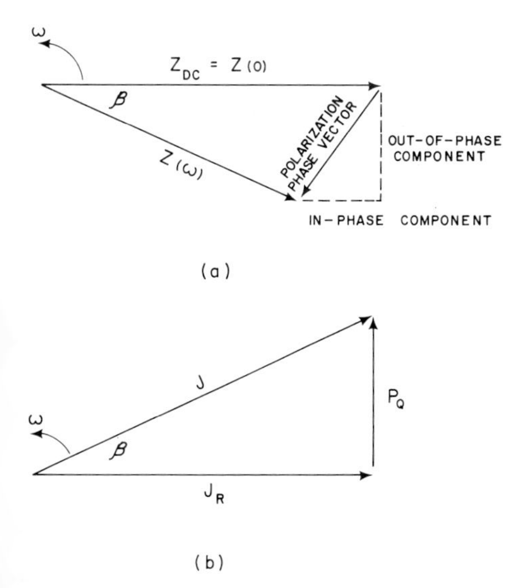

Fig.3.23. Phasor diagrams.

frequency spectrum will show a phase-angle maximum corresponding with the steepest part of the Warburg impedance region (Fig.3.24).

The metal factor can be viewed as the out-of-phase conductivity of a polarizable material where phase is the angle whose tangent is the ratio of out-of-phase to in-phase components.

Mathematical formulation of IP response

In the following development, it must be remembered that the differential of a logarithmically varying quantity is really only the fractional amount of change in its value; that is:

$$\mathrm{d} \log \rho = \mathrm{d}\rho / \rho$$

Also, from eq. 7, it is helpful to think of chargeability simply in terms of a change in resistivity:

$$M = \Delta \rho_t / \rho$$

so that:

$$M = \mathrm{d} \log \rho \tag{60}$$

With eq. 60, the earth's linear polarization phenomena can be treated mathematically.

Equations for determining the apparent chargeability M_a for any subsurface polarizable body and any electrode arrangement have been presented by Seigel (1959b). The total apparent chargeability M_a is a summation of all chargeabilities M_i times the weighting function B_i, or $M_a = \sum_i^n M_i B_i$. The summation of all weighting functions is one, or $\sum B_i = 1$. The general expression for a subsurface composed of n different materials of resistivities ρ_i, where $i = 1, 2, 3, \ldots, n$, is:

$$M_a = \sum_i M_i \frac{\partial \log \rho_a}{\partial \log \rho_i} \tag{61}$$

which is to say that the apparent chargeability of the various polarizable components in a subsurface system are represented by the ratio $B_i = \partial \log \rho_a / \partial \log \rho_i$, or to the ratio of change of apparent resistivity (polarization) to the resistivity of the particular component. A family of dimensionless weighting B_i functions $\partial \log \rho_a / \partial \log \rho_i$ can be plotted as type curves for specific examples of subsurface geometries and surveying arrays.

In the frequency measurement method, the function $\partial \log \rho_a / \partial \log f$ would apply, where f is frequency and:

$$\frac{\partial \log \rho_a}{\partial \log f} = \sum_i \frac{\partial \log \rho_a}{\partial \log \rho_i} \frac{\partial \log \rho_i}{\partial \log f} \tag{62}$$

The characteristic property of the individual ratio $\partial \log \rho_i / \partial \log f$ is represented by the observed variation with frequency in the same proportion as was

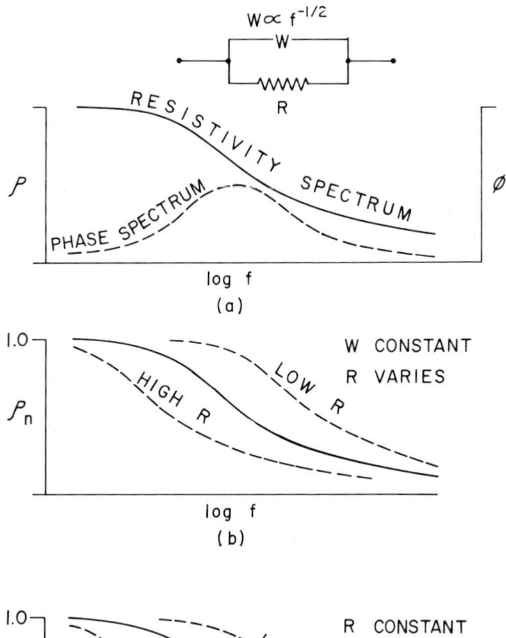

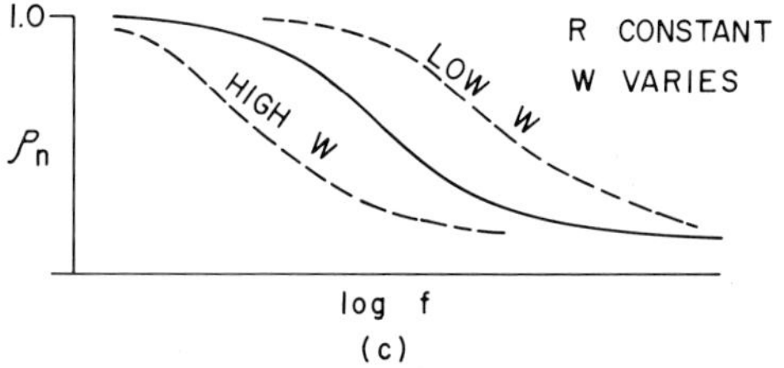

Fig.3.24. Frequency characteristics of a parallel circuit in which the Warburg impedance (W) and the resistance (R) vary.

M_i, or $\partial \log \rho_a / \partial \log \rho_i$. Thus, chargeability can be transformed to frequency by:

$$M \rightarrow \frac{\partial \log \rho}{\partial \log f}$$

Now:

$$\frac{\partial \log \rho_a}{\partial \log f} = \frac{1}{\rho} \frac{\partial \rho}{\log f}$$

and one can keep the $\partial \log f$ (or $\Delta f/f$) interval constant by measuring IP effects at 0.1, 0.3, 1, 3 Hz. Then, M is proportional to $(1/\rho_f)\ (\rho_{f_2} - \rho_{f_1})$; that is, M is proportional to the slope of the resistivity spectrum normalized to some resistivity ρ_f (which could be ρ_{f_1}) against $\log f$. This relationship was indicated in Fig.1.5.

For a single polarizable region in an otherwise nonpolarizable earth (Fig.3.25):

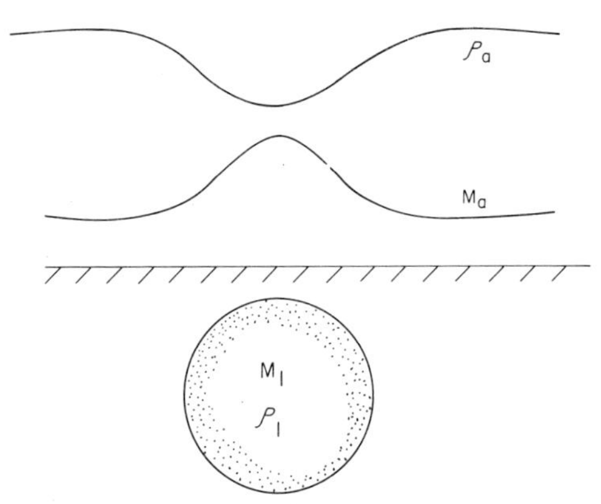

Fig.3.25. Response of a polarizable sphere.

$$M_a = M_1 \frac{\partial \log \rho_a}{\partial \log \rho_1}$$

and:

$$M_a = M_1 \rho_1 \frac{1}{\rho_a} \frac{\partial \rho_a}{\partial \rho_1} \tag{63}$$

For the frequency method:

$$(FE)_a = \frac{(\rho_{dc})_a - (\rho_{ac})_a}{(\rho_{ac})_a} = \frac{\Delta \rho_a}{(\rho_{ac})_a}$$

and:

$$(FE)_a = \frac{1}{\rho_a} \frac{d\rho_a}{df} \Delta f$$

Since only ρ_1 is dependent on frequency:

$$(FE)_a = \frac{1}{\rho_a} \frac{\partial \rho_a}{\partial \rho_1} \frac{d\rho_1}{df} \Delta f$$

$$= \left(\frac{1}{\rho_1} \frac{d\rho_1}{df} \Delta f\right) \rho_1 \frac{1}{\rho_a} \frac{\partial \rho_a}{\partial \rho_1}$$

Since:

$$\frac{1}{\rho_1} \frac{d\rho_1}{df} f = \frac{\Delta \rho_1}{\rho_1} = (FE)_1$$

$$(FE)_a = (FE)_1 \rho_1 \frac{1}{\rho_a} \frac{\partial \rho_a}{\partial \rho_1} \tag{64}$$

Because the maximum true frequency effect $(FE)_1$ in the polarizable medium is approximately equal to the maximum chargeability M_1, comparison of eqs. 64 and 63 shows that the measured apparent frequency effect

$(FE)_a$ is approximately equal to the apparent chargeability M_a. Therefore, either the apparent frequency effect or apparent chargeability can be calculated if the apparent resistivity is known as a function of ρ_1.

Magnetic induced polarization

In 1974, Seigel announced a new IP technique, which he named magnetic induced polarization (MIP). Magnetic fields associated with polarization decay are measured with a sensitive magnetometer. The method works best near elongate polarizable bodies. As yet, it is only used in special projects of a research nature.

THE EQUIVALENCE OF FREQUENCY AND PULSE MEASUREMENTS

There is sometimes discussion on the relative merits of the frequency and pulse IP methods. Insofar as the field instruments and the sometimes ingrained field techniques are concerned, there can be a definite difference between these two methods, but they are theoretically equivalent. This has been previously pointed out in this book as well as by Madden and Marshall (1958), Wait (1959), and Hallof (1964).

The basic requirement of frequency- and time-domain equivalence for electrical circuits is the linearity of the passive system being studied. In other words, the response frequency of the output must be the same as that of the input. Then a Fourier or Laplace transform is applicable from one coordinate system to the other.

For a material with a specific impedance $\rho(j\omega)$, Ohm's law for a three-dimensional material that has a complex resistivity is:

$$E(j\omega)\, e^{j\omega t} = \rho(j\omega)\, J\, e^{j\omega t} \tag{65}$$

where E is electric field intensity and J is current density, as before. The limit values of the complex resistivity are:

$$\lim_{\omega \to 0} \rho\,(j\omega) = \rho_0 \quad \text{and} \quad \lim_{\omega \to \infty} \rho\,(j\omega) = \rho_\infty$$

Frequency effect can be defined as:

$$FE = \frac{\rho_0 - \rho_\infty}{\rho_\infty} = \frac{\rho_0}{\rho_\infty} - 1 \tag{66}$$

Therefore:

$$\rho_0 = \rho_\infty\,(1 + FE)$$

In the time domain, current is applied in the form of a step function, and the decreasing voltage response function $V(t)$ is a diminishing transient, as was indicated in Fig.1.1. If a substitution $s = j$ is made, Laplace transform theory dictates that:

$$\mathscr{L}\,[V(t)] = V(s)\,\frac{1}{s} = \frac{\rho(s)}{s}\,J \tag{67}$$

From the initial and final value theorem for Laplace transform theory:

$$\lim_{t \to \infty} V(t) = \lim_{\omega \to 0} J\,\rho(j\omega) = J\rho_0$$

and:

$$\lim_{t \to 0} V(t) = \lim_{\omega \to \infty} J\,\rho(j\omega) = J\rho \tag{68}$$

A definition of chargeability M for a homogeneous polarizable material responding to a step function current is:

$$M = \frac{\underset{t \to \infty}{V(t)} - \underset{t \to 0}{V(t)}}{\underset{t \to \infty}{V(t)}}$$

This defining equation is the same as eq. 55, $M = V_s/V_p$.

Thus, from eqs. 67 and 68:

$$M = \frac{J\rho_0 - J\rho}{J\rho_0} = \frac{FE}{1 + FE} \tag{69}$$

and for small frequency effects where $FE \ll 1$:

$$M \approx FE \tag{70}$$

Note that if the frequency effect was defined $FE = (\rho_0 - \rho_\infty)/\rho_0$ instead of as in eq. 66, the correlation of FE and M would be exact rather than the approximation given in eq. 70.

The above derivation draws a parallel between high-frequency and dc responses and initial and final values of pulse transient voltages. At intermediate frequencies or transient times, as displayed by decay curves and resistivity spectra, the correlation is more complex.

THE PLACE OF IP THEORY IN THE EXPLORATION METHOD

For the explorationist, a rigorous understanding of the theory of induced polarization is not an absolute necessity. One can apply basic concepts of the first-order IP effect and find sulfide bodies. It can also be said that a complete founding in IP theory is unnecessary and that theory can only be an idealized approximation to environmental fact.

Although these statements are in a sense geologically true, they should not discourage the geophysicist and the inquiring explorationist from continuing to investigate the interesting and important details of IP theory.

Chapter 4

LABORATORY WORK IN INDUCED POLARIZATION

This chapter will be concerned with laboratory studies of the IP effect and with related electrochemical findings that are pertinent to understanding of this phenomenon. Laboratory observations are important because here environmental conditions can be closely controlled and measured. IP behavior of a rock specimen can be observed in detail, as well as some of its physical properties that give rise to the directly measured IP effect. However, great care must be taken to assure that only natural IP phenomenologic effects are being measured in the laboratory procedure rather than artifacts of the measuring system or exterior extraneous effects.

In particular, the basic causes of induced polarization can be investigated under controlled conditions as a function of current density, ion type and concentration, pore geometry, or temperature and pressure. Mineralized rock can be studied with variation and variety of metallic mineral and particle size and shape. Also, the IP phenomenon in anisotropic and unmineralized rocks is important. It is readily possible to review the response of rocks in a given area or of a given type in terms of metallic mineral content or as related to decay-curve characteristics.

If possible, it is advisable to obtain and measure electrical properties of a suite of rock samples from an area before it is surveyed in order to learn the IP characteristics. With this information, interpretation of field data can be more precise because there are a smaller number of unknowns in attempting to understand the subsurface. Exploration interpretation problems stemming from field conditions in which there might be a high IP background can be anticipated before uncertainties arise. It is logical that samples from the first drill holes be tested for their IP characteristics.

Because of the disturbing effects of sampling procedures on the sample and the occasional uncertainties of some laboratory measurements, it is often desirable to make in-place resistivity and IP measurements. Drill-hole IP measurements are sometimes made for in-place test purposes. Although it is not always easy to do so, electrically sampled in-place material should be positively identified by every practical means available. If possible, short electrode intervals should be used in drill-hole surveys because of the uncertainty of property changes with depth. Here also, current density and sometimes electrode effects must be definitely considered. One can concisely compare high-quality laboratory test work with carefully made in-place field measurements.

After an area is surveyed and drilled, it sometimes becomes important to be certain the explored IP anomaly was indeed explained by the findings of the drill. Laboratory IP tests on questionable samples and perhaps in-hole surveys can provide a final answer to this important exploration question. Visual inspection of a core or a representative rock sample will frequently lead to identification of the polarizable component. However, a visual analysis alone may not predict whether a sample is polarizable or to what extent.

These are some of the reasons for making small-scale, in-place IP measurements for studying rock samples in the laboratory and for having an appreciation and understanding of the significance of laboratory-derived results.

ELECTRICAL MEASUREMENT OF ROCKS

Physically, a rock is a complex composite of minerals, each with independent structural and compositional characteristics. In addition to the solid components of rocks, the arrangement and geometry of the voids are important, as well as the composition of fluids and gases in the rock. From an electrical standpoint, the pore spaces in earth materials and their shapes are every bit as important as the solid constituents.

Although the metallic-luster minerals in a rock are often in the form of veinlets rather than being truly disseminated, sulfide grains are usually discontinuous and are commonly associated with pore openings along which the sulfides were deposited by mineralizing solutions. Observation of the minute textural details of a mineralized rock will impress the careful observer with the extent and role of the tiny fractures. A sulfide-bearing rock that has been enriched by the supergene process is illustrative of the importance of the nature of pore space. Pores must be present, but their presence is not immediately visually obvious. Unfortunately, the important pore-geometry problem in mineralized rocks has not received much detailed attention from geologists and geophysicists.

A rock sample will not contain all the fractures of the sampled region that are so important electrically. Fractures are difficult to collect, but their effect is one of the variables to be considered before applying the results of laboratory studies to field situations. Comparison of field resistivity results with laboratory sample values from the same area shows that resistivities taken in place are lower than those obtained from samples, pointing out the nature of the resistivity sampling problem. The low resistivity of the water-filled fractures is not represented in the samples. But IP values taken in the laboratory are much more like those taken in place, as might be expected in these sampling conditions.

There is sometimes a danger in extending all of the IP features detected in synthetic samples to natural samples. Synthetic rock may be necessary for some special laboratory studies, but in general, there is no guarantee that the

IP mechanisms and effects observed in these substances are the same as those found in natural materials.

Types of experiments

The metallic mineral particles in a rock are complex natural electrodes whose properties may be studied either in place or within artificial surroundings. IP experiments on laboratory samples can be controlled for study in several ways. Electrolyte composition and degree of saturation may be changed, also temperature and pressure. Observed changes can be correlated with theoretical models.

It is good laboratory practice to have one or more 'control specimens', whose polarization and resistivity properties will depend on the type of experiment to be performed. Repeated comparative measurements assure calibration of the laboratory system in terms of known value standards, such as an unpolarizable sandstone or a moderately polarizable sulfide specimen. Of course, the properties of the standard must be carefully preserved.

Results of laboratory experiments on the behavior of polarizable rock under high current density conditions can be applied to the interpretation of field problems where polarizing rock occurs near the current electrodes. These experiments can explain the sometimes peculiar in-place IP results in circumstances where mineralized rock is near an electrode.

Synthetic rock materials can be created for specific purposes, their advantage being control of otherwise complexly varying natural conditions. Nevertheless, a synthetic rock is only an approach to a model, and the chosen variables unfortunately may not be the important ones.

The goal of IP laboratory experiments is to attain a better understanding of specific polarization properties of rocks and also to learn more about general polarization characteristics. In this way, field data can be better interpreted.

Sampling and preparation

Guidelines that are ordinarily used for sampling geologic materials hold for the selection of laboratory samples. 'Specimens' may be sought as being typical of minerals, crystal habit, etc., whereas a 'sample' should be representative of the rock type being mapped or the rock involved in the exploration program. For purposes of interpreting data, a representative sample is required, and a sampling procedure often calls for some geologic sleuthing. Unoxidized samples without unusual through-cutting veinlets of sulfides or gangue minerals are of course preferable. Hopefully, the sample should be large enough that a cube of useful size can be cut from it so that smaller inhomogeneities will not contribute to a measurement. Current densities are easier to control at the necessary low levels on larger pieces, and there is less evaporation of natural pore fluids both immediately after collection and during measurement. Larger samples have a greater volume to surface area ratio so that cut surfaces of sulfide particles will contribute less of an undesirable surface polarization effect. Drill core with trimmed ends and sawed to lengths about the same as the core diameter makes good laboratory samples.

Oriented samples of foliated or otherwise anisotropic rocks are desirable for the interpretation of some field measurements. Symbols can be made for a north arrow, dip, and sample number with an indelible marker. Sample locations, of course, should be described and marked on a map which also has survey line locations, geology, and assay information. The sample should be put immediately into a water-tight container in order to preserve its natural pore fluid. Clean plastic sacks are usually quite satisfactory, while metal containers can be suspect.

Laboratory preparation of the sample involves cutting a cube or parallelepiped with a large-diameter saw. Faces of cubes should be marked for reference, particularly with foliated rocks. Samples of other shapes may be prepared, as from cylindrical drill core, but it is often desirable to make azimuthal measurements on the rock as can best be done from rectangular solid shapes. Water with the same composition and conductivity as the rock's original pore fluid is preferable for cutting fluid. Tap water is not the most efficient lubricating coolant, and furthermore, a prompt and thorough cleaning of the cutting enclosure is necessary to prevent rusting of the saw's parts. Cathodic protection can help preserve the saw. Rock pore fluid or its substitute is often corrosive.

If a water-soluble oil or detergent is used as a cutting fluid, its conductivity should be determined because it can be assumed that this fluid will penetrate the sample to some extent because of temperature, pressure, and activity differences. A controlled experiment using two similar but differently sawed samples of a porous material, such as a brick or a homogeneous sedimentary rock, will prove the effect of a contaminant. It is usually quite difficult to remove such material from a rock.

After cutting, the sample should be stored in water, preferably of the same composition and conductivity as the pore fluid. The water should be deoxygenized, and for some purposes it can be de-ionized. The cutting fragments should probably be saved for other physical property measurements.

In the laboratory, the electrical measurement is made in an enclosure or a humidifying chamber to minimize evaporation from the surface of the specimen. It is not desirable to coat or cover the surface of a sample in an attempt to avoid evaporation unless great care is taken not to create a conductive path. However, it is sometimes convenient for some experiments to cast the ends of the sample permanently in appropriately shaped plastic holders. Reactive casting material should be avoided.

If the sample has dried out, water can be reintroduced by placing the rock in a vacuum chamber and then immersing the rock under a vacuum or by soaking it for an extended period, more than several days, while periodically measuring its electrical parameters. Membrane polarization materials, such as clay minerals, frequently lose most of their polarization properties after drying or heating and reintroduction of pore fluid. It is possible to make resistivity measurements while the sample is being soaked and to observe that the resistivity asymptotically approaches a stabilized value in a few days. As a matter of fact, resistivity measurements made during the reintroduction of pore fluid are an excellent method of evaluating the sample's permeability. Of course, the porosity of the sample can be found by subtracting the dry from wet weight and dividing this figure by the volume of the sample.

Loose material, such as rock chips or cuttings, can be tested by placing it in a specially constructed sample container of specified geometry with built-in electrodes. The measured apparent IP effect can even be larger than the true value of in-place polarizable material because of the creation of additional surface area. The additional interstitial water lowers the resistivity and also the volume polarization.

Electrical measuring equipment

By present standards, many of the early pioneering IP measurements were rather primitive, mainly because of the disadvantages of older, vacuum-tube type test equipment. The current switching was usually done mechanically or manually, which created bothersome electrical transients. The sensitivity of current-regulating and voltage-measuring equipment was not very high, so that fairly high current densities were necessary to get accurate measurements. It is now recognized that current densities above 10^{-3} A/m^2 (0.1 μA/cm^2) give rise to nonlinear, current-saturated IP responses with the amount of both electrode and membrane polarization lessening. Scott and West (1969), for example, find that their synthetic rock samples give only a linear response below $1.25 \cdot 10^{-3}$ A/m^2. Nevertheless, the earlier experimenters made many meaningful measurements, and the present improved systems are based on those of the past.

The modern instruments available for laboratory IP measurements have undergone improvements similar to those found in contemporary field equipment. Solid-state preamplifiers with an input impedance of more than $10^{10}\ \Omega$ minimize polarization effects of receiver electrodes and accompanying error in resistivity measurements.

The current source used for laboratory experiments described in this chapter was especially designed and built for the purpose of making sensitive IP measurements at the Geophysics Laboratory, The University of Arizona. It is a solid-state, constant-current waveform generator and preamplifier. Output is well calibrated, can be monitored, and is continuously adjustable from 1 μA to 11 mA at frequencies from dc to 1,100 Hz. Duty time is continuously variable between zero and 100%, giving waveforms ranging from spikes of alternating polarity to a true full square wave. The input on the high input-impedance preamplifier is differential, with gains of 1, 10, and 100, and the input has a dc voltage buckout to counteract any dc potentials in the electrode or sample system. The output of the preamplifier is not differential but is single-ended relative to chassis ground, which greatly simplifies the connections to chart recorders.

The amplified decay voltage signal can be monitored on a chart recorder or photographed on an oscilloscope. For frequency variation studies, the voltage signal can be passed through a narrow band-pass filter to examine the amplitude of the fundamental harmonic sine wave, which can then be displayed on a voltmeter or oscilloscope. Phase-angle measurements can also be made.

Laboratory IP work discussed in this chapter is an extension of earlier experiments reported by Madden and Marshall (1958), Fraser et al. (1964), McEuen et al. (1959) and Collett (1959), all of whom laid ground work for present techniques. Most of the experiments described in this chapter were done by K.L. Zonge and W.A. Sauck at the Geophysics Laboratory, the University of Arizona, using natural electrolyte with a resistivity of 20 Ωm and a current density less than 10^{-2} and 10^{-3} A/m^2, except where so indicated. This work is being further investigated by Katsube and Collett (1973).

Environmental chamber

It is desirable to carry out laboratory measurements in an enclosing humidifying environmental chamber because drying of the surfaces of a sample will continually raise its apparent resistivity. This type of chamber can be evacuated or heated for other types of electrical rock property measurements. It also acts as an electrostatic shield when grounded to reduce the electrical noise that can otherwise be bothersome to the low voltages being measured. A photograph of a vacuum-oven environmental chamber containing sample holders is shown on Fig.4.1.

The resistivity of rocks decreases with temperature, and the relationship between the resistivity of pore water and temperature is given as:

$$\rho_T = \rho_{20}\ e^{-0.022\,(T-20)}$$

where ρ_T is the resistivity of water at temperature T and ρ_{20} is the resistivity of water at 20°C. This temperature effect amounts to a total resistivity change of about —2%/°C, so that some form of temperature stabilization is not only desirable but for porous rocks rather necessary. The thermal variation of induced polarization has been predicted, but it is not well established by published results, although Collett (1959) noted an increase in membrane polarization and a decrease in electrode polarization with increasing temperature. If constant standard temperature cannot be maintained, it is well to note and stabilize a temperature of 20—25°C in the measuring chamber.

Fig. 4.1. An environmental chamber used in testing IP properties of rocks.

Sample holders

A circuit diagram for a sample undergoing measurement can be quite simple, the circuit diagram portion of the sample-holder portion of Fig.4.2 being almost the same as that of Fig.2.4. There have been a number of different kinds of sample holders used by various investigators. Some of the basic features of these holders merit some discussion.

The simplest sample testing arrangement, which is a fairly effective one for making only a few basic measurements, consists of sandwiched felt pads and metal-gauze electrodes at either end of the rock sample, as illustrated in Fig. 4.3. If there is a low current density and high input impedance measuring system, the material for the metal electrodes can be copper, brass, or even

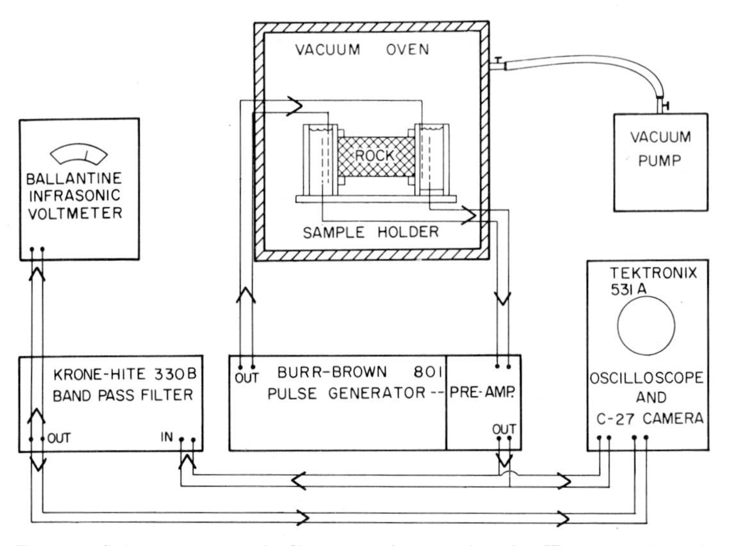

Fig.4.2. Schematic circuit diagram of a test for the IP properties of a sample.

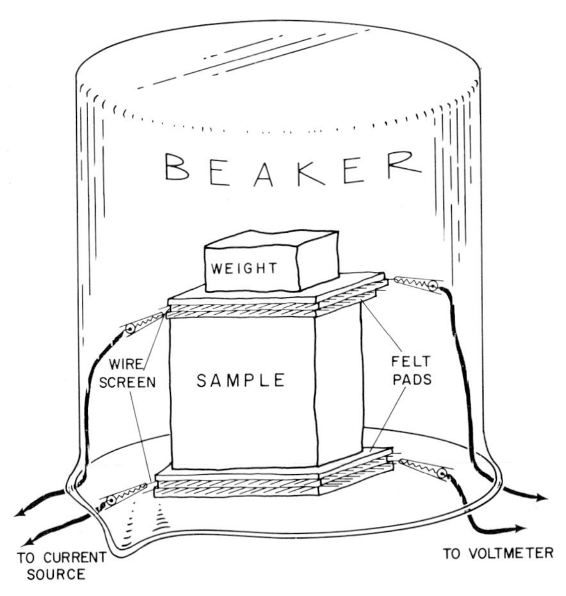

Fig.4.3. Simple sample testing arrangement.

galvanized iron screen with only a low correctable difference in polarization effect. In the past, silver and silver chloride electrodes have also been used to reduce the polarization and noise that may arise at the screens.

Most investigators prefer to have the potential electrodes on the outside of

Fig.4.4. A specially designed sample holder for use in an environmental chamber.

the electrode-sample system. This is to avoid passing current through the measuring electrodes and so prevent any possible polarization of the potential contacts. A weak salt solution in the felt-pad electrode contacts has sometimes been suggested to reduce the resistance at the electrodes, but it may also lower the measured resistivity, and it will in time change the ion properties of the sample. The sample and associated electrodes can be placed under a humidifying cover, such as an inverted beaker, to complete this basic sample holder.

Polarization will take place between exposed metallic mineral faces and the contacting electrolyte, and this artificially created polarization effect may become an appreciable fraction of the total polarization. Even the presence of bubbles at the contact interface may cause spurious polarization.

Some readers may feel that it is unnecessary to go into 'cookbook' details for apparently simple resistivity and IP measurements, but in this writer's experience, basic concepts are not always intuitively achieved. For example, several years ago a mining company engaged a well-recognized research organization to make laboratory IP measurements so that the company could objectively gauge the effectiveness of the IP method. The laboratory insisted on using dry rock samples with dry metal-to-rock electrode contacts. Then the laboratory, after working for one year and spending a considerable sum of money, reported to the mining company that the IP method would yield inconsistent results if used in prospecting for mineral deposits. Not only was time and money wasted, but the company and the research laboratory were retarded in their development of the IP technique as a worthwhile exploration tool.

The sample holder shown in Fig.4.4 is made of cemented and machined Plexiglas and holds four samples in horizontal position between pairs of small electrolyte tanks, each of which holds a current and a potential electrode near the face of the sample. The current electrode is nearest the sample. Wire leads are short to reduce resistive and capacitive effects. For these reasons, the preamplifier was mounted directly on the sample holder. Plastic collars or flanges are cast around the ends of each sample; therefore, the entire end areas of the sample are in contact with the electrolyte, allowing unimpeded resistivity measurements to be made. To calibrate electrodes and to measure precisely electrolyte resistivity, a water-filled Plexiglas tube can be sampled in the holder between electrolyte reservoirs. A good sample holder should be capable of extended use for periods of days or weeks under a variety of testing conditions, and it should give repeatable results.

FREQUENCY-DOMAIN STUDIES

The full square wave type of waveform used in frequency-domain studies of the IP effect is illustrated in the top part of Fig.4.5. The time axis is horizontal, increasing to the right, and current is displayed on the vertical axis. The lower part of Fig.4.5 shows the pulsed square wave of the time-domain current waveform for which current is switched 'off' twice during the cycle in a (+, 0, —, 0, +) sequence.

One way of displaying the change in resistivity with frequency is shown on Fig.4.6, where the upper photograph shows successive uniform decreases in voltage (resistivity) as the frequency is increased by factors of ten. This photograph was taken as five separate exposures, with the oscilloscope sweep rate being increased by a factor of ten. The percent frequency effect (PFE) is measured as the percent difference in peak amplitude between the signals at any pair of frequencies. Note that this difference, which is due to the IP effect, is about the same for every increment of frequency for the entire

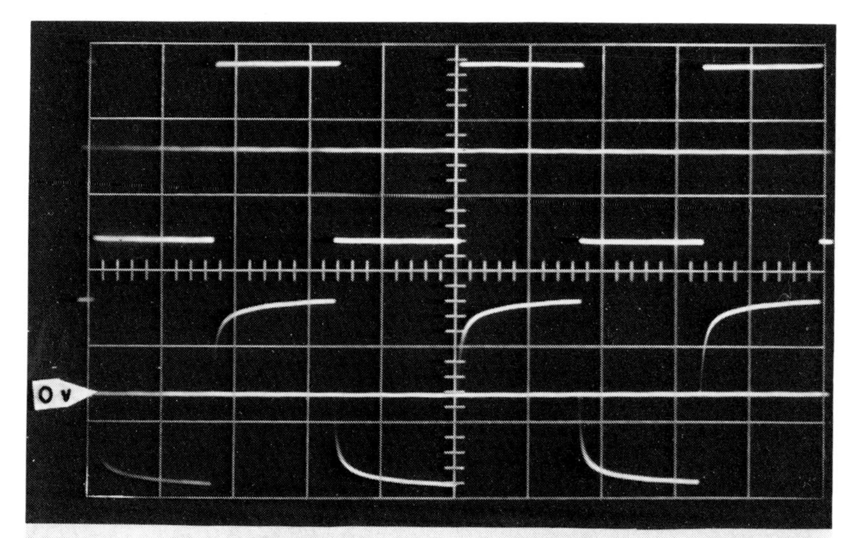

UPPER—CURRENT WAVEFORM FOR FREQUENCY DOMAIN.
LOWER—VOLTAGE ACROSS POLARIZABLE SAMPLE.

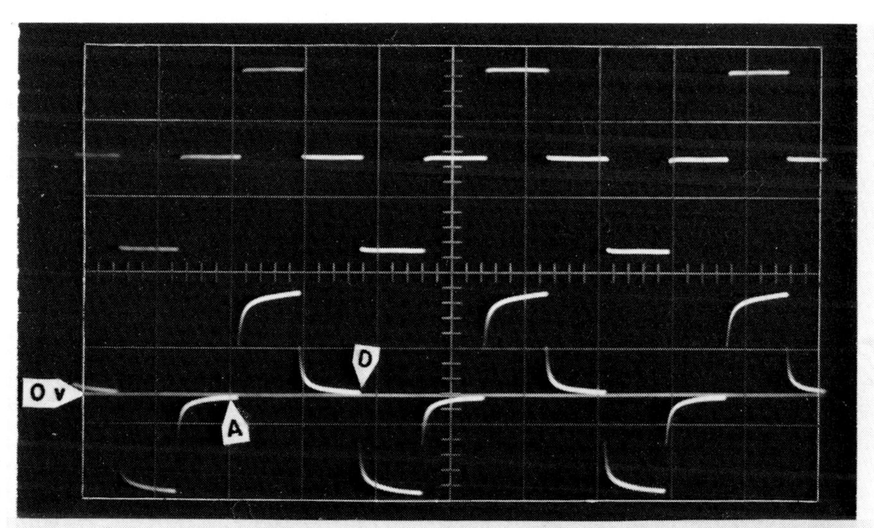

UPPER—CURRENT WAVEFORM FOR TIME-DOMAIN MEASUREMENTS.
LOWER—VOLTAGE ACROSS POLARIZABLE SAMPLE.

Fig.4.5. Frequency-domain (upper photograph) and time-domain (lower photograph) waveforms, displaying current as a function of time.

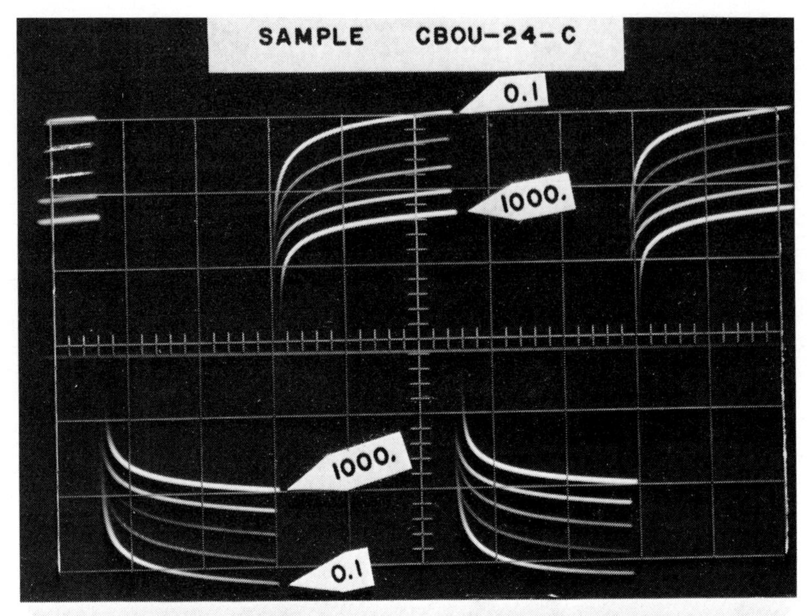

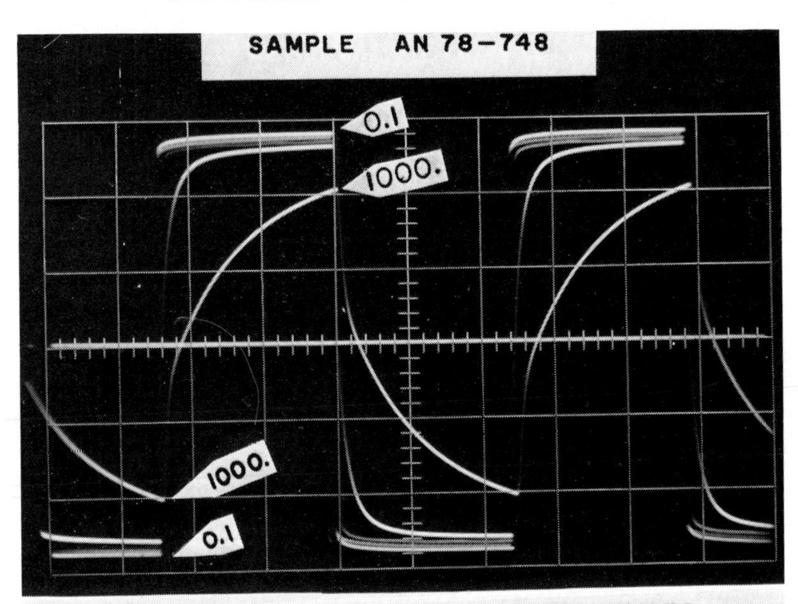

Fig.4.6. Resistivity responses with frequency for two samples. Upper photograph of an oscilloscope recording shows IP effects linear with frequency for five frequencies. Lower photograph shows IP effects nonlinear with frequency for the same five frequencies.

range of frequencies shown. Thus it could be said that the voltage difference or IP response of this sample is geometric, or logarithmic, with time, and the response is therefore fairly constant over each decade in this large frequency range. This sample is from a highly graphitic shear zone and does contain a scattering of microscopic pyrite, although less than 1% by volume.

By contrast, the lower part of Fig.4.6 shows a different picture. Here there is little frequency effect at lower frequencies but a large difference between the 100 and 1,000 Hz traces. The horizontal line at the center marks zero volts. It can be said that this sample has a nongeometric IP response with time over successive decades in this frequency range. This sample is quite impermeable and is a low-porosity silicified rock with visible pyrite blebs making up almost 1% of the rock volume.

RESISTIVITY SPECTRUM

A resistivity spectrum is a display of rock resistivity as a function of frequency. To construct such a diagram the resistivity or voltage level is assigned 100% at the datum frequency. Resistivities measured at other frequencies over the spectral range are normalized relative to this datum. For example, a frequency of 0.03, 1.0, or 1,000 Hz could be selected as a base value. We have chosen the low-frequency voltage value as the normalizing reference. This reference value is analogous to normalizing the time-domain voltage V_s of eq. 55 to have a value of 1.0, or 100%.

The frequency given on a resistivity spectrum is presented on a logarithmic scale, which conveniently displays decade (factor of ten) frequency intervals as equal intervals, similar to Fig.1.5. If the plotted function is nearly logarithmic, it becomes a nearly straight line. A logarithmic scale condenses and expands values appropriately. Also the frequency-scale reading error $\Delta f/f$ is constant over the scale length.

A resistivity spectrum is the frequency-domain counterpart of the IP decay curve. The resistivity change $\Delta\rho_f$ of the resistivity spectrum curve is a measure of the IP response. If this interval is normalized by dividing by the resistivity, one obtains the normalized slope, which is the IP frequency effect.

In making IP measurements on laboratory specimens over a standard one-decade frequency difference, one does not encounter $f^{-1/2}$ frequency effect values that are as large as the maximum theoretical percent. Instead a maximum dependence of $f^{-1/4}$ is to be expected from a distributed circuit; this effect amounts to 78% over one decade. IP frequency effect values that are this high have only occasionally been approached and are seldom exceeded, except in massive sulfides which have artificial surface polarization or in anisotropic rocks.

Interpretation of the resistivity spectrum

One can view the resistivity spectrum of a polarizable rock as a representation of the Warburg impedance, related to the rock's resistive and non-faradaic impedances, as was illustrated in Fig.3.7. The resistivity spectrum of a rock is concave down at the low-frequency end of the Warburg impedance and convex down at the high-frequency end. This relationship was noted by Fraser et al. (1964), who then pointed out that massively mineralized rocks should have convex-down spectral curves. Disseminated mineralized or membrane polarization rock with higher resistivity should ideally display concave-down spectral curves. This relationship holds fairly well for most mineralized rocks, with occasional exceptions, such as shown on Fig.4.7.

The resistivity spectra shown in this chapter are normalized to the low frequency. This display method changes the curve shape somewhat from that of plotted conductivity spectra normalized to the high frequency.

Effects of current density

Figure 4.7 is a typical resistivity spectrum showing the effect of increasing current density from 10^{-3} to 10^{-2} A/m^2. Resistivities of this particular sample are normalized at 0.03 Hz, but they differ by 10% at the highest frequency. The IP effects of the lower frequencies used in the field can therefore be

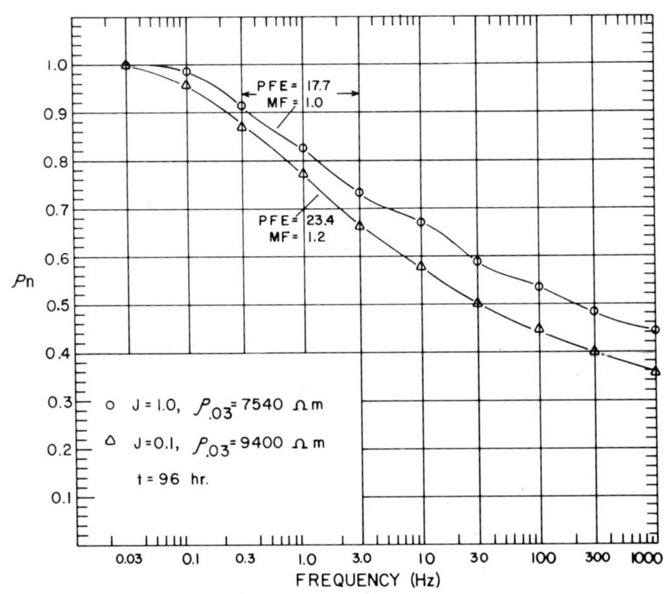

Fig.4.7. Resistivity spectra for sample HC-1605B, showing a slightly concave-up portion of the curve. Spectra for disseminated samples often display a markedly convex-up shape.

expected to decrease with an increase in current density above 10^{-3} A/m^2. This sample contains a single, longitudinal, one-mm wide, somewhat discontinuous veinlet of pyrite, and there is no metallic mineralization exterior to the vein.

Figure 4.8 shows similar trends of decreasing resistivity and decreasing IP with an increase in current density. This phenomenon is due to current saturation. The PFE is halved in going from 10^{-2} to 1 A/m^2. There is also a small change in spectral curve shape with current density.

Another way of displaying these kinds of data is illustrated in Fig.4.9. The resistivities of all four spectra are normalized to the 0.03-Hz value, using 10^{-2} A/m^2. The resistivity shown by the lower curve is seen to be 1.46 times that of the upper, at the high frequency. The sample is a black, hard, fine-grained metamorphic sedimentary rock containing graphite and very fine grained pyrite visible only under high magnification.

The electrical properties of rock thus change drastically with change in current density. IP effects are greatest and are linear with current density at low current densities and, as a general rule, fall off markedly as densities increase.

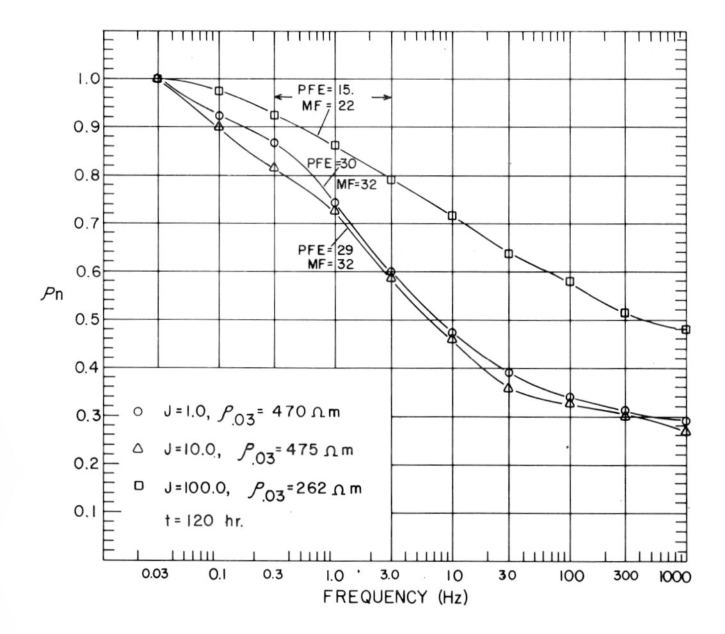

Fig.4.8. Resistivity spectra showing effects of varying current density. Units of current density are given in μA/cm^2. Sample CBOU-51-B.

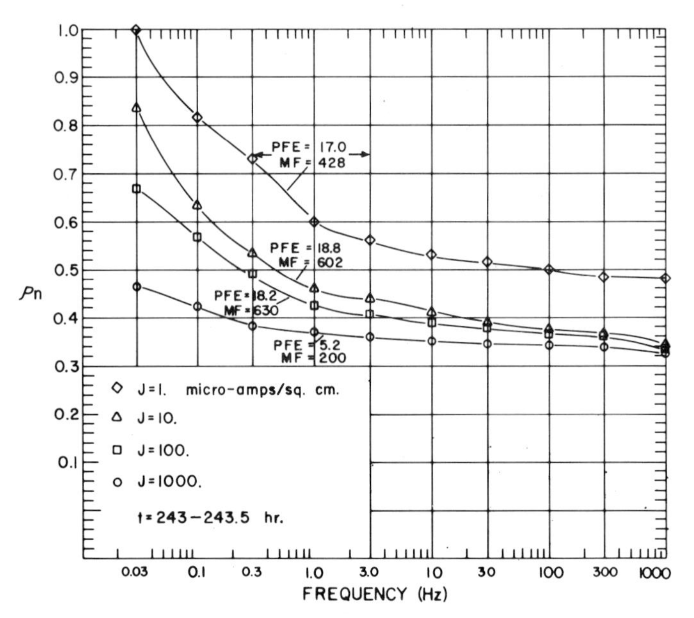

Fig.4.9. Current density variation effects on the resistivity spectra of sample Minn-1.

Effect of rock saturation

In another experiment with the sample shown in Fig.4.9, a constant current density was used with IP measurements taken after various soaking times, as shown on Fig.4.10. It is evident that the low-frequency resistivity increases with sample saturation in apparent violation of Archie's law. This sample has a resistivity even lower than that of its electrolyte, a situation which is occasionally encountered in mineralized rocks. The cause of the increase in resistivity with soaking time is probably due to the blocking action of the more mobile diffuse-layer ions. Here we see (Fig.4.10) a series of convex-down spectral curves that Fraser et al. (1964) have reported as being related to massive mineralization. Although this sample is not massive in a mineralogic sense, it behaves as such from an electrical-property standpoint.

Figure 4.11 shows another example of variation of resistivity with degree of fluid saturation. This is an illustration of a concave-down curve. The IP effect at field frequencies is very low, but it is much greater at frequencies above 30 Hz. Note that the high-frequency IP effect decreases as the sample becomes more saturated with electrolyte. This implies that some partially saturated rocks can give greater IP effects than saturated ones. This is probably due to conductivity variations of the diffuse layer as affected by pore geometry. This sample from the Homestake formation near Lead, South Dakota, is primarily a cummingtonite schist with chlorite and biotite and with minor sulfide mineralization.

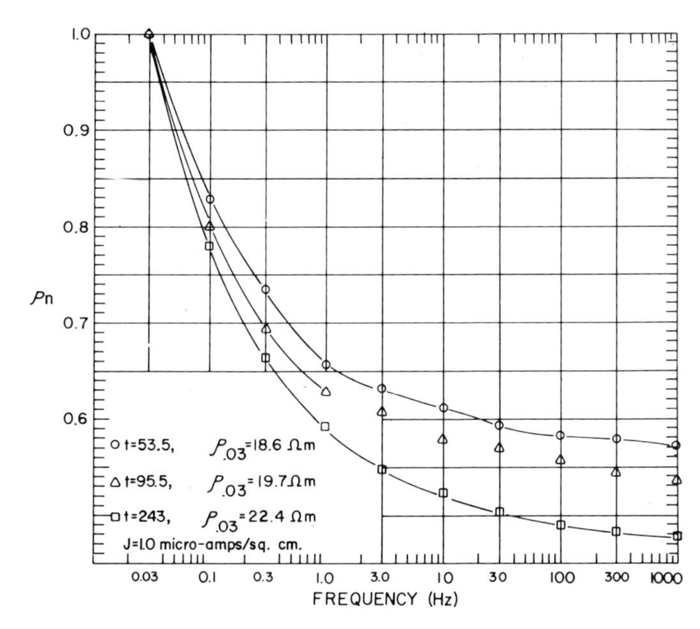

Fig.4.10. Normalized resistivity spectra for various soaking times. Sample Minn-1.

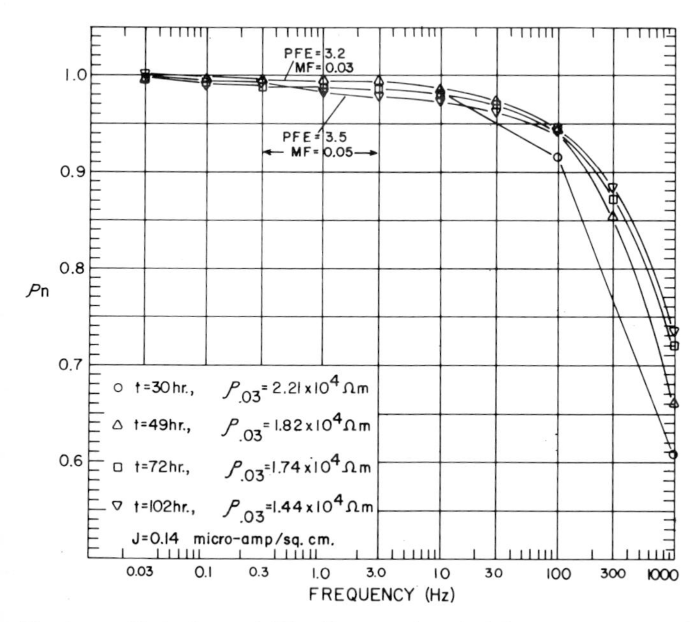

Fig.4.11. Variation of IP effect and resistivity spectra as a function of soaking time. Sample HS-756.

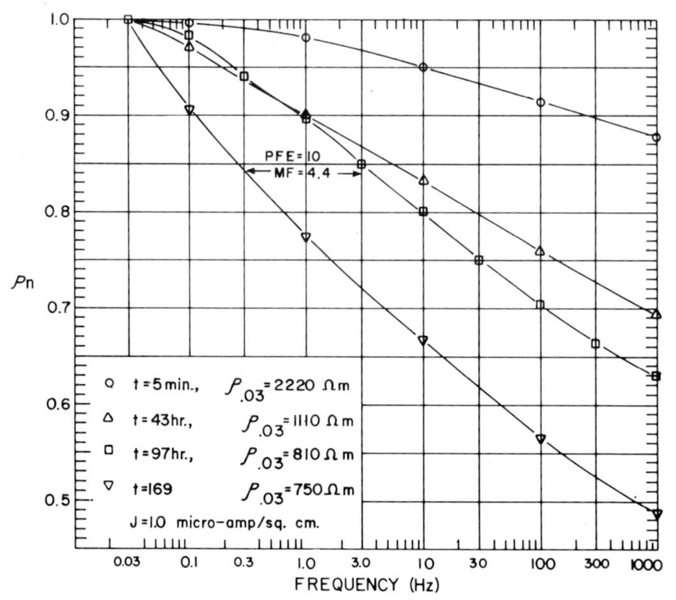

Fig.4.12. Normalized resistivity spectra for sample CBOU-24C, showing a marked concave-down shape.

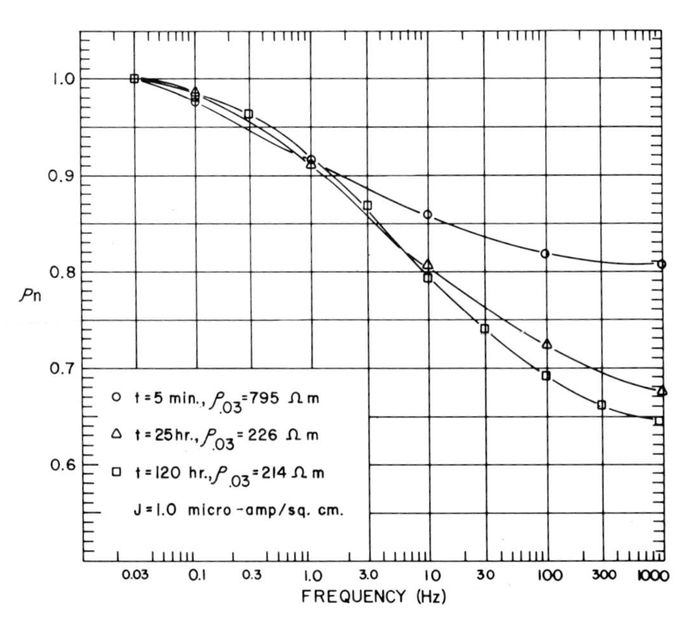

Fig.4.13. Normalized resistivity spectra for sample CBOU-51-B, showing effect of increasing pore-fluid saturation.

The sample data shown in Fig.4.12 display a more conventional effect of rock saturation. The slope of the resistivity spectral curves, which is the IP response, increases with electrolyte content. The resistivity spectra are nearly linear. The data for this sample, showing waveshape change with frequency, are compatible with the results of Fig.4.6 for the same sample.

The shape of the resistivity spectral curve can also change with the degree of saturation of a sample. Figure 4.13 illustrates this variation, which may have practical significance in field operations in a circumstance where one may tend to exaggerate the importance and meaning of the resistivity spectrum.

ANISOTROPISM

It has been pointed out that foliated rocks have a lower true resistivity in a direction parallel to the plane of foliation. Kunetz (1966) discusses the coefficient of anisotropy λ, defined as $\lambda = \sqrt{\rho_t/\rho_L}$ where ρ_t and ρ_L are the true resistivities in the transverse and longitudinal directions, respectively. In most foliated rocks, the coefficient λ is less than two.

The true IP effect in anisotropic rocks, measured either as chargeability or frequency effect, is greater in the direction parallel to the plane of foliation than when measured across it. However, the percent difference of IP values is not as great as the percent difference of resistivities.

Figure 4.14 shows spectral curves of a specimen of schist from the Homestake mine in which λ is 14.5, which is exceptionally high. The PFE ratio between 0.3 and 3.0 Hz, however, is 4.7:1, or about one-third of the coefficient of anisotropy and 1/44 of the resistivity ratio. Note that the contrast of true metal-factor measurements in the two directions is 279:1, which is absurdly high and must be due to the difference in pore geometry with direction in the specimen. The meaning of the metal-factor parameter is therefore much less certain in anisotropic rocks than in isotropic rocks.

Resistivity and IP effects can be measured in the laboratory as functions of soaking time, as is also shown on Fig.4.15. Therefore, it is possible to study indirectly the relative permeability in the two directions, which would be otherwise difficult to measure.

Generally, resistivity varies with permeability, as it does with porosity. The rate of change of resistivity with soaking time is greater for permeable rocks than for impermeable ones. Figure 4.16 shows how the resistivities of six widely differing samples, dry at the start of the experiment, change as the samples become saturated with electrolyte. Soaking times have been divided by sample lengths for normalization. Time elapsed at inflection points marked *a* through *f* can be considered as inversely proportional to fluid permeability. Except for BCOU-33-B, which had a high graphite content, samples with higher resistivities have lower permeabilities. PFE and MF also increase with permeability.

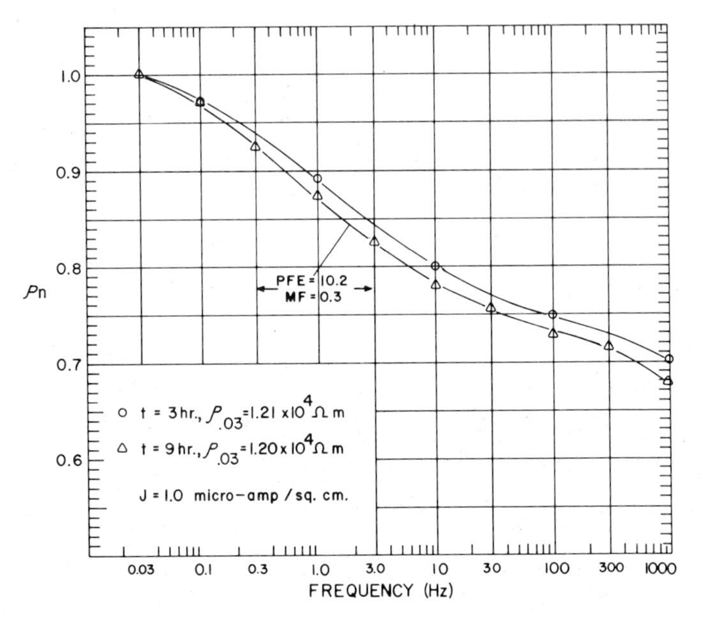

Fig.4.14. Normalized resistivity spectra for sample BC-F.

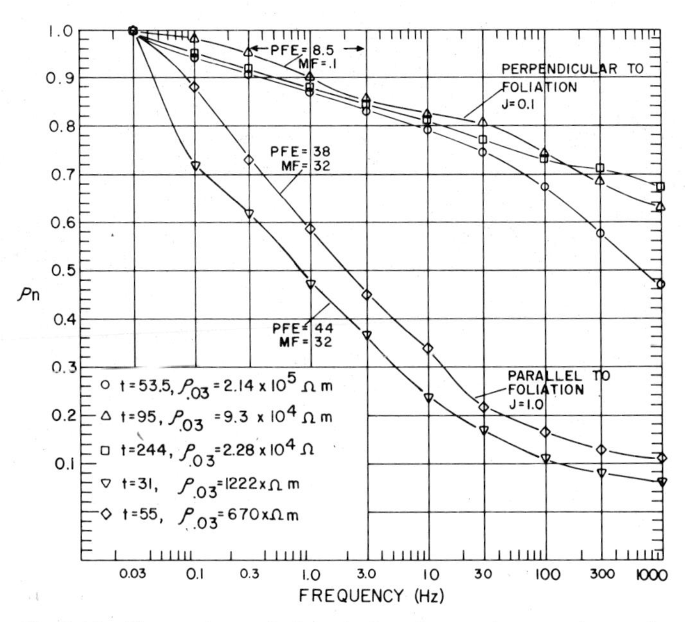

Fig.4.15. Illustration of electrical anisotropism as shown by the resistivity spectra of sample HS-13.

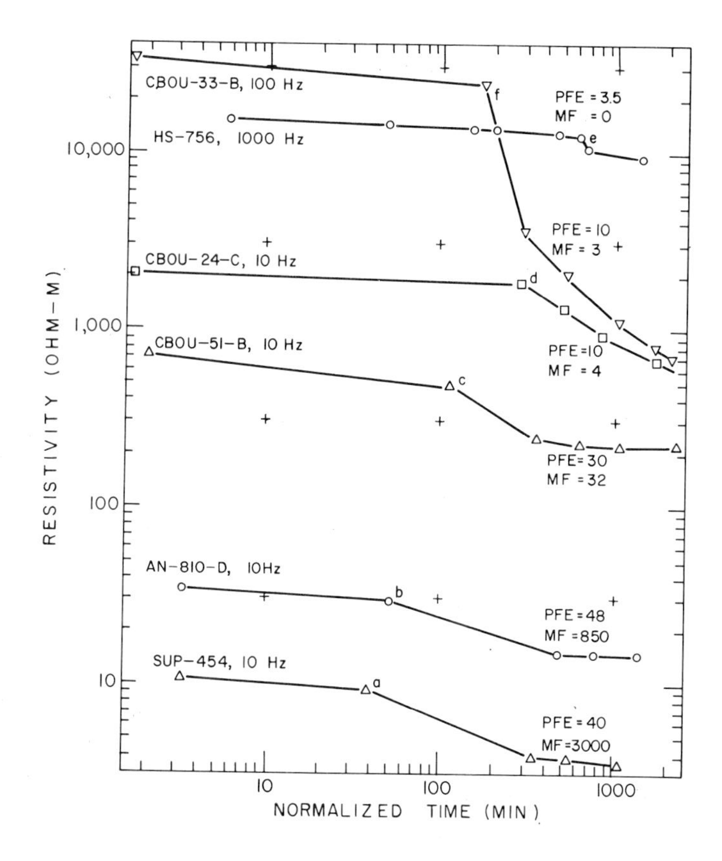

Fig.4.16. Variation of resistivity with the duration of time that a sample is soaked with water.

TIME-DOMAIN MEASUREMENTS

There are several different ways of measuring the IP effect in the time domain; the method is versatile. One measure is the area (commonly in millivolt-seconds) between the decay curve and the zero volt level. This value is usually normalized by dividing it by the peak charging voltage. Chargeability M is fundamentally defined as the ratio of the initial decay voltage to the maximum voltage of the charging cycle. Referring to Fig.1.1, chargeability is $M = V_s/V_p$, where V_s is secondary voltage and V_p is primary voltage. Another important parameter of the decay curve is the exponential time constant τ, defined in the conventional fashion as the time interval from the current 'off' instant to the time when the voltage level is 1/3 or 37% of its initial value, as shown on Fig.4.17.

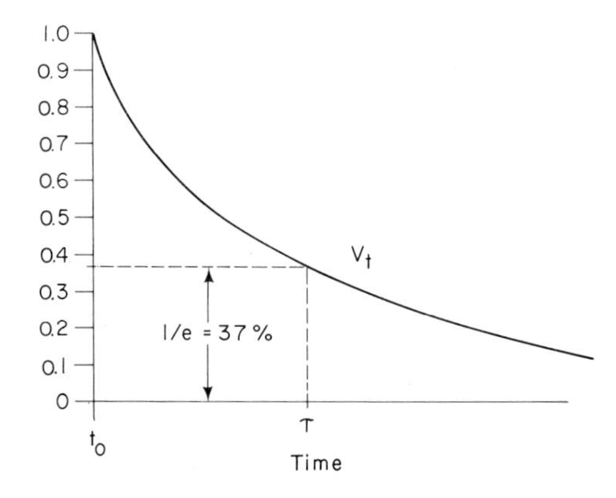

Fig.4.17. Illustration of the time constant τ of a decay curve.

Charge and decay cycles

Oscilloscope photographs are useful in studying short-period decay curves because the electron beam has practically no inertia as compared with a pen recorder system, the instrument has a very high input impedance, and the time scale can be expanded for display purposes by making successive sweeps of the beam. On the other hand, wide chart recorders are convenient for collecting longer period IP data.

For very long charging times (several minutes), the transient voltage across the sample is higher, giving larger, longer lasting decay voltages. This longer time system, advocated by Bertin (1968) and Compagnie Générale de Géophysique, gives better resolution for detailed measurements of the longer period IP effect. Decay voltages in polarizable earth materials will persist for times longer than the length of the charging cycle.

Figure 4.18 shows the charging curve and decay curve plotted on a common logarithmic time scale. By plotting both curves in this way, any difference between change and decay can be noted, as is done on the nearly straight difference plot at the bottom of the diagram. This diagram, similar to Fig.1.5, uses real data.

The shapes of logarithmically plotted decay curves differ from one sample or area to the next in much the same way that resistivity spectra differ from one another. These time-derived curves can be roughly transformed to frequency curves when desirable. A decay curve contains considerably more information than the area under the curve, or frequency-effect measurements.

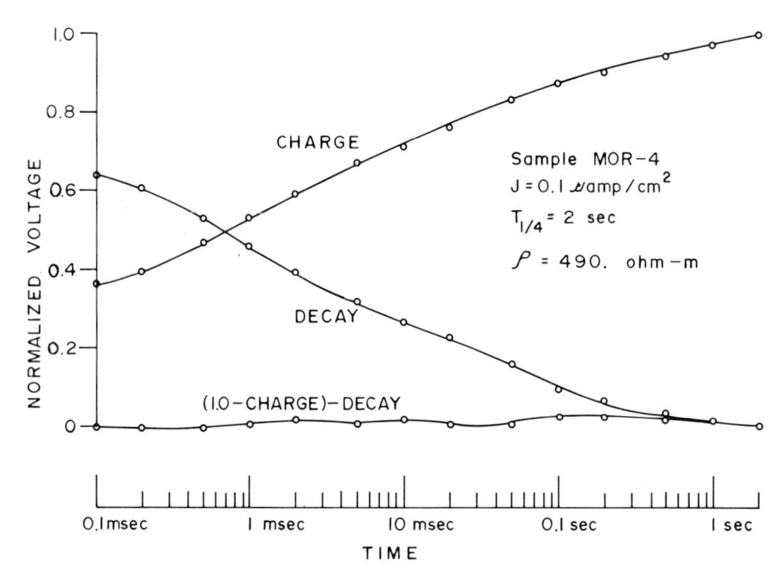

Fig.4.18. Charge and decay curves plotted logarithmically with time. A nearly straight line difference curve at the bottom of the diagram shows the linear nature of the IP phenomenon.

PHASE MEASUREMENT

A phase spectral plot of resistivity values can be made in much the same way as a frequency spectrum. Figure 4.19 shows such a diagram that compares both the resistivity and phase spectra of a laboratory sample. The phase angle is proportional to the slope of the resistivity spectrum curve at low values of phase angle. Phase angle is proportional to the derivative of the resistivity spectrum curve. To carry this comparison to larger phase angles, note that resistivity is a complex quantity with in-phase and out-of-phase components. Phase angle β is the angle whose tangent is the ratio of these two values, which can be written:

$$\beta = \tan^{-1} \frac{\rho_{\text{imag}}}{\rho_{\text{real}}}$$

The voltage amplitude on a phase-angle spectral plot, such as Fig.4.19, is the vector sum of the in-phase and out-of-phase components of resistivity. The maximum phase angle is found at the inflection of the resistivity spectrum curve.

Phase-angle measurements are not particularly difficult to make over large frequency variations, either in the laboratory or in drill holes. The limit of the phase spectrum as well as the resistivity spectrum is more restricted in routine field surveying than in the laboratory or in drill holes.

The equivalent of a $\Delta\rho/\rho$ determination is obtained with a single

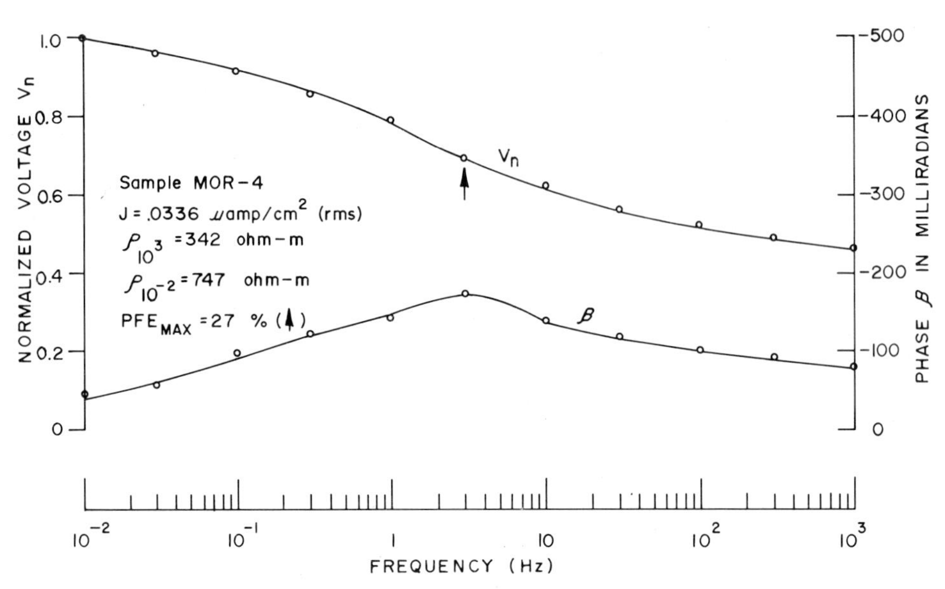

Fig.4.19. Comparison of a phase-angle spectrum and a resistivity spectrum.

measurement of phase rather than two frequency or time measurements required with frequency or pulse methods. Phase-angle information can also be a useful supplement to resistivity spectral measurements in the field in reducing the problem of electromagnetic coupling.

EARLY PART OF DECAY CURVE AND THE COMPARISON WITH FREQUENCY EFFECTS

The relationship between time-domain and frequency-domain measurements can be recognized. A sample with a small IP effect at high frequencies and a much larger effect at low frequencies will show little short-period decay but more long-period decay.

The end of the charging voltage (*A* and *B*) and the very early part of the decay curve are shown on Fig.4.20. A zero-volt line is indicated near the bottom of the photograph. Current is the same for each pair of traces. Indicated frequencies are those of the current waveform. The charging time is one-fourth of the period of the waveform; that is, the charging time at 0.3 is 100 times as long as that for 30 Hz.

Sample P-44 shows only a background level of polarization, and the sample is effectively depolarized after less than half a millisecond. The sample contains 80% wollastonite and 20% calcite.

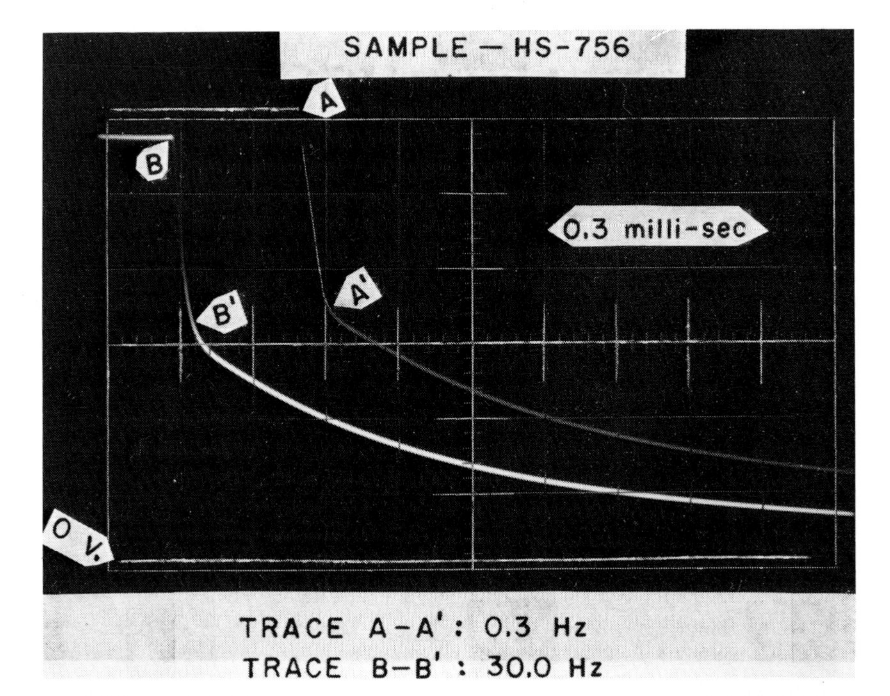

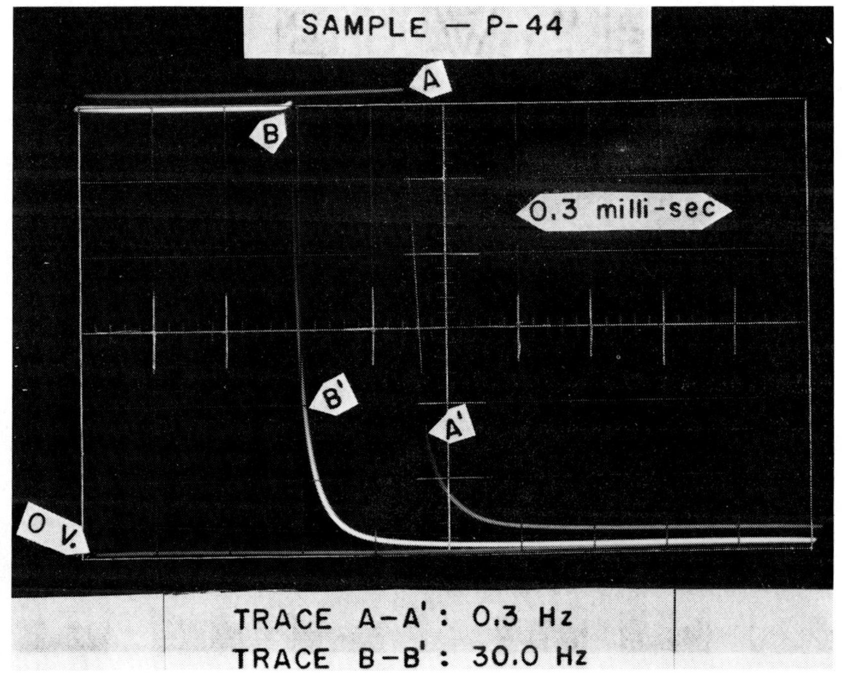

Fig.4.20. Rapid sweep rate photographs of the early parts of the decay waveforms of poorly and moderately polarizable samples.

Sample HS-756 shows polarization lasting for only one to two milliseconds with only little longer time decay. In transforming to the frequency domain, one would then expect only a small low-frequency IP effect but some high-frequency effect. This is borne out by the frequency data of the sample, as shown on Fig.4.11, where the maximum PFE occurs between 300 and 1,000 Hz (periods of 3.3 and 1.0 milliseconds, respectively).

CONSIDERATIONS OF LABORATORY AND FIELD RESULTS

Comparison of frequency-domain laboratory IP results with field results is meaningful. Any major lack of correlation is probably due to too large current densities used in the laboratory or to geometric peculiarities of the polarizable body in the field. Figure 4.21 shows several typical resistivity spectral curves that were shown earlier in this chapter but are replotted as IP 'spectral' (actually cumulative IP curves), using Hallof's (1967) linear frequency scale. Because data plotted in this linear frequency fashion will always give concave-down curves, this plotting method will not appreciably differentiate between polarization types. The logarithmic frequency scale gives better resolution to low-frequency phenomena and better discrimination between different spectral types.

Data shown in this chapter are sometimes inconclusive, but it is evident that laboratory studies of induced polarization help bridge the gap between IP theory and applications in the field. Additional detailed laboratory studies and correlations of IP data will no doubt be fruitful.

One could categorize the measured magnitude of IP response into first-

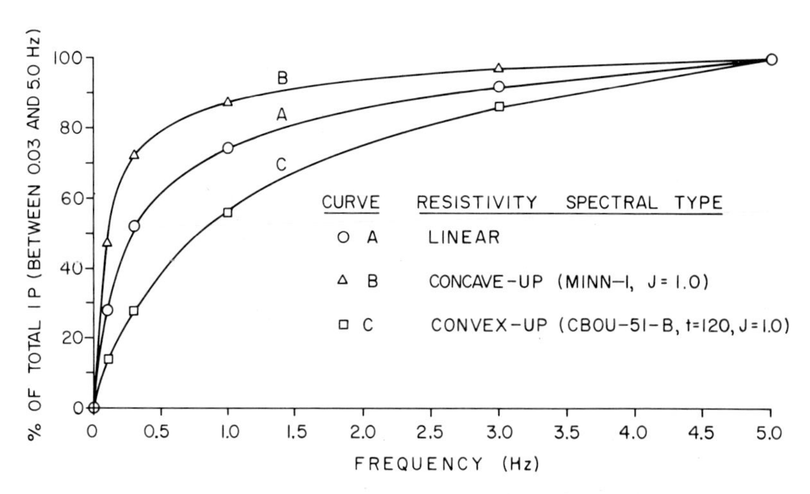

Fig.4.21. Typical resistivity spectral curves replotted on a linear scale, after the method of Hallof (1967).

order and second-order effects. A first-order effect would be the basic manifestation of the IP phenomenon, expressed as chargeability, percent frequency effect, or metal factor. The shape of an IP decay curve or a resistivity spectrum would be a second-order expression. The second-order effects are seen and measured in laboratory experiments, where the many variables of the environment can be studied and controlled. The physical cause and meaning of the more subtle expressions of induced polarization seem to be important clues to a more complete understanding of the details of the phenomenon.

Chapter 5

IP FIELD EQUIPMENT

One of the initial appeals of the IP method is its apparent simplicity in making and understanding field measurements. The IP method is somewhat unique in exploration in its presentation of an anomalism against a practically zero-level background, ideally giving easily interpretable results in a 'yes—no' form. This apparent simplicity is attractive to the electronic engineer designing field equipment as well as to the exploration manager trying to comprehend and evaluate field results. The matter of field instruments is actually more involved than appears at first impression.

It is not too difficult to put a piece of mineralized rock into a demonstration tank, to insert metal electrodes in the water, and with standard laboratory electronic test equipment to show some important features of the IP phenomenon. However, there are difficulties in putting similar instruments into the field where power and sensitivity requirements are taxed to the limit and where specific conditions within the broad range of environment often differ from idealized laboratory design specifications. It thus becomes necessary to build the sensitive electrical measuring devices with field surveying requirements and conditions in mind. This chapter will discuss the various requirements of field equipment and the ways in which designers have met and solved problems.

HISTORICAL DEVELOPMENT

The practical importance of suitable field measuring apparatus is emphasized by the fact that the IP effect was first observed empirically by Schlumberger (1920), using resistivity measuring equipment rather than being hypothesized from laboratory work or theory. Observation of the IP phenomenon was verified much later by an electrolytic tank experiment by Brant in 1946, who was working with equipment developed by coinvestigators at the Radio Frequency Laboratories in Boonton, New Jersey. The state-of-the-art at that time is summarized in Brant and Gilbert's Patent No. 2,611,004 (1952). Seigel's (1949) and Ruddock's (1959) descriptions of early field instruments illustrate the simplicity and effectiveness of their applied experiments. In 1960, Newmont Exploration Limited supported a program for IP instrument design with Varian Associates of Palo Alto, California, for the development of low-frequency, phase-sensitive, frequency-domain equipment. Comparative field tests with this equipment have demonstrated the equivalence of pulse, frequency, and phase-angle IP measurements. Dolan and McLaughlin (1967) have summarized this work and also their more recent time-domain IP instrument development.

Bleil's (1953) pulse IP equipment conceived during World War II and the equipment of Vacquier et al. (1957) were similar to Newmont's. The first commercially available pulse IP equipment was marketed by Hunting in 1957 and was followed by that of Seigel and associates in 1958.

Madden and his students at MIT used their frequency IP equipment in a field survey in Nova Scotia in 1955 (Madden and Marshall, 1958). Advantages in the frequency equipment design inspired McPhar in 1957, Geoscience, Inc. in 1959, and Heinrichs Geoexploration Co. in 1960 to develop their own apparatuses. Mining companies, including Kennecott and Phelps Dodge, produced similar frequency-domain instruments, while ASARCO adhered to the pulse equipment. Anaconda's geophysical division under McAlister perfected phase-sensitive field equipment in 1955.

DESIGN CONSIDERATIONS

Basically there are two separate circuit systems in all IP field equipment: a signal-generating and current-transmitting complex and a voltage-receiving instrument with its related circuits. Although these two systems are usually operationally independent, during measurement they are necessarily coupled through the intervening earth.

Sometimes an IP transmitter and a receiver have been packaged together and their output combined to provide a single resultant reading. Also, the two units are sometimes physically and electrically time linked so that a more precise phase reference is provided for the received signal. But it is advantageous to be able to separate the two units physically when carrying out most of the IP field surveys. The intended use of IP equipment, whether for routine exploration surveys in a specified range of environmental conditions or for experimental purposes under a variety of conditions, is a matter for initial consideration. Also, the designer must bear in mind the convenience of the field operator and his limitations and capabilities for servicing and repairing.

Parameters to be measured

The electrical nature of the earth, particularly its polarization properties, is an important consideration in the design of an IP field system. The electrical earth acts in both a passive and an active fashion, with the sought-for passive characteristics being measured through the interference of the active noisy part. The type of three-dimensional electrical measurements that must be made have been previously discussed with the conclusion that four contact points (Fig.2.8) are necessary for the two circuits. The electrode layout and field procedures are also an important consideration, which will be considered in a later chapter.

Electrode resistance

The term 'electrode resistance' requires explanation. Of course, the metal electrode material itself offers virtually no electrical resistance, neither does the wetted, disturbed zone adjacent to the electrode. But the undisturbed earth immediately beyond strongly contributes to the total resistance of the grounded circuit, and it is this zone that contributes to the electrode resistance, R_E. The resistance range of the grounded contacts is of initial importance. Both the current and potential electrode resistances are controllable variables. The importance of lower resistance warrants spending time and effort in preparing the electrode sites. In Figure 5.1, a generalized relationship of the electrode and ground resistances, particularly of a current electrode circuit, is shown as a function of the amount of work necessary to lower this resistance. Note that doubling the labor in making electrode contacts will not always halve the resistance but that it is necessary to expend a certain amount of labor to make an acceptable electrical contact. Usually, a resistance level of from 100 to 1,000 Ω can be obtained without undue effort, depending on local soil conditions. Rocky or sandy areas have high surface resistances, and multiple parallel-connected electrode sites are necessary in order to allow sufficient current to be passed through the transmitting circuit to receive a strong voltage signal. It is not possible to measure the resistance R_E of a contact of a single electrode with the ground without reference to another electrode.

As in all information communicating systems, a high signal-to-noise ratio

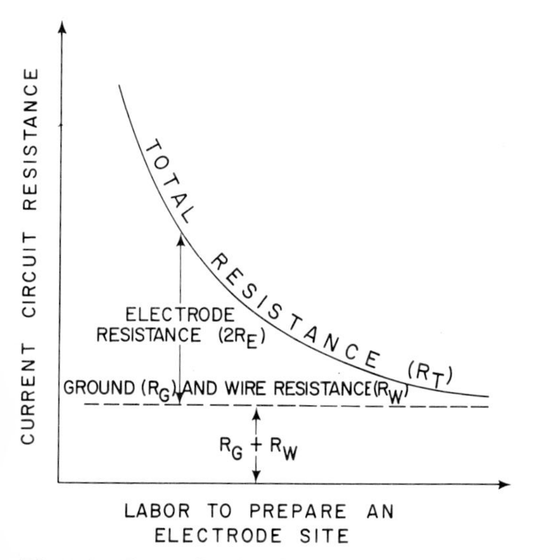

Fig.5.1. Generalized relationship between the labor necessary to prepare an electrode site and resistance of current circuit.

is necessary to make low-error measurements. The lower the resistance of the contacts, the less dissipation of the total amount of available power. Therefore, maximum current can be sent and a larger voltage signal received by making the resistance of the current electrode contacts with the ground low.

An idealized circuit showing electrode resistance as they relate to other circuit values is shown in Fig.5.2. The total current circuit resistance R_T is:

$$R_T = R_W + 2R_E + R_G \tag{71}$$

where R_W is the wire resistance and R_G is the ground resistance.

By using a four-electrode system, only the ground resistance (impedance) itself couples the transmitter and receiver circuits (Fig.2.21), and this resistance is, of course, the important geophysical quantity to be measured in the field. However, the other resistive quantities shown in Fig.5.2 enter into the determination of an optimum behavior for individual transmitter and receiver circuits and they certainly must be taken into account in the overall design of field instruments.

Nonresistive impedance effects at the current electrodes due to electrochemical polarization are usually not large enough to affect IP measurement of ground properties. These effects can be minimized by using a constant-current type IP transmitter. Current-electrode impedances can be readily checked by a field calibration method, as will be discussed in the chapter on field survey methods.

Potentials generated by chemical reactions of the current-electrode materials may cause the resistance of the ground circuit to vary considerably, depending on the direction of measurement (polarity). This slight 'diode effect', which has also been called 'polarization', at the current electrodes causes an additional complexity in a frequency-method transmitter by requiring it to send a constant current of alternating polarities into the ground. However, the asymmetry between the measured forward and back electrode resistances usually decreases as current is cyclically reversed through the ground circuit.

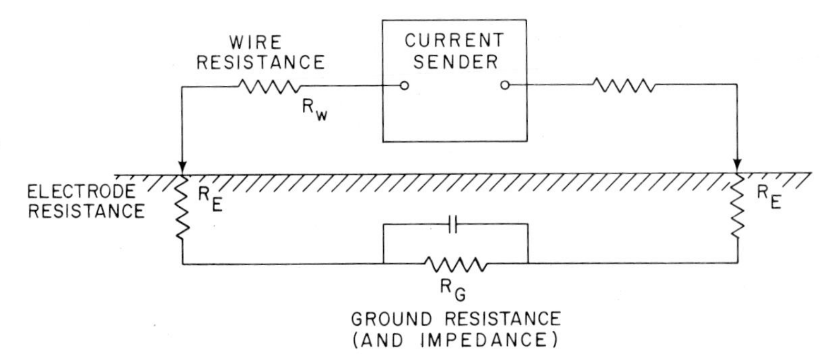

Fig.5.2. Resistances of the current electrode circuit. The two electrode resistances R_{E_1} and R_{E_2} may not be equal and may differ with the direction of current flow.

Since conventional pulse IP transmitters must provide a given amount of 'off time' in their total duty cycle, the motor generator may surge unless appropriate precautions are taken. A compensating dummy load or an automatic throttle will prevent surging.

After an optimum expenditure of effort in preparing the electrodes (Fig.5.1), the electrode resistance R_E is reduced. The undisturbed zone near the electrode, within 5% of total interelectrode distance, contributes almost all the electrode resistance. The many possible current paths in the zone between the electrodes tend to minimize the contribution of the ground resistance effect, even though the resistance along each parallel path is rather high. The electrode resistance R_E and the ground resistance R_G are directly related to resistivity. The wire resistance R_W is, of course, controllable and should be made smaller than the electrode resistances but within the reasonable bounds set by cost, tensile strength, and weight of wire. Typical wire resistances would range from 2 to 50 Ω per thousand feet, depending on requirements of the survey.

The transmitter designer sometimes views his load specifications as being within the range of the probable field conditions shown in Fig.5.3, where the current circuit resistance can vary between high and low limits. Since a power source can deliver $I^2 R$ watts into a resistive load R, the generalized resistive load characteristics of a current transmitting system can usually be described by the type of curve shown in Fig.5.3. Most of the total current circuit resistance R_T is due to the electrode resistances, but some is due to the ground resistance. There is an obvious advantage in utilizing all of the electric

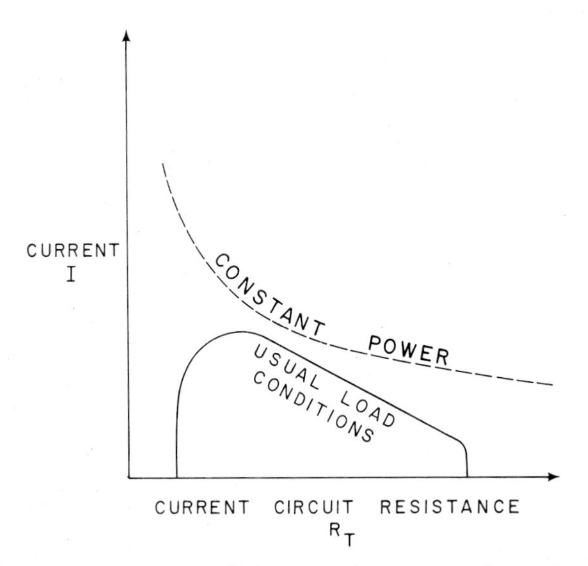

Fig.5.3. Load conditions of the current electrode circuit showing current I and total resistance R_T. Larger currents can be attained with more effort expended in preparing electrode positions.

power available to put the most current into the ground, but this maximum is usually a trade off with the amount of labor that must be expended to obtain a low electrode resistance. Also, it is not easy to design a transmitter that can always take advantage of a low-resistance ground circuit because of the current and voltage limitations imposed by other design features.

If one is dealing with both a fixed electrode resistance and a maximum available power output level, the maximum current that can be put into the ground is fixed at:

$$I_{max} = \frac{V_{max}}{R_T} = \sqrt{\frac{\text{Power}}{R_T}}$$

subject to the maximum ratings of the components of the system. The transmitting system must be primarily a regulated current source, and for this reason a variable output voltage must be available in order to put the desired current through a given resistive load R in accordance with Ohm's law, $V/I = R$. Here again, current-sending devices are not always easy to optimize for maximum or near-maximum current delivery. To allow for any increase in electrode resistance due to drying out of electrodes or other causes, the practical maximum current is 5—10% less than the available maximum. In any event, the designer must assume that field personnel will realize and utilize the advantages by providing the lowest possible stable electrode resistance.

Voltage and polarization signals

In the grounded potential measuring circuit, the potential electrode resistance can vary between a few hundred and perhaps as much as 100,000 Ω without causing serious inaccuracies in readings at a voltmeter-receiver. Potential electrodes with abnormally high resistances are undesirable because they make the voltmeter-receiver more susceptible to the noise signals that can sometimes be traced to poor electrical connections or transient voltages. If the voltmeter does not have a much higher input impedance than the electrode resistance by perhaps a factor of more than 100 to 1, current drawn from the ground may cause voltage changes in the ground itself, causing measurement errors.

Although it may be desirable at times to measure IP effects in different frequency bands or intervals, generally the received frequency range in the field will be below 10 Hz in either pulse or frequency methods. There is no arbitrary lower frequency limit to resistivity and IP signals, but sometimes the deletion of very low frequency voltages is helpful in reducing telluric or self-potential noise. Duration of measuring time is a factor in field efficiency that the equipment designer must also take into account: the lower the frequency, the longer the time to take a reading.

The voltage level of received primary voltage signals V_p measured in the field can be as low as 10 μV. It is desirable to have the precision of a measurement made by the frequency method as high as 0.1% but to achieve a comparable IP reading accuracy in the pulse method, a receiver need have a sensitivity of only about 2.0% in measuring the decay voltage. These precisions may not be easily obtainable at very low signal levels, especially under noisy field conditions.

The necessary high sensitivity of IP equipment and the needed precision of measurement are major differences between IP and resistivity field instruments. The maximum received potentials can be between 1 and 10 V, but the most commonly used voltage range would probably be between 1 and 10 mV, depending on a number of prevailing field survey factors.

Various characteristics of the IP decay curve have been studied by different groups, but there are not many general statistical compilations of the basic phenomenon beyond observing that to a first approximation the decay curve is logarithmic and the IP phenomenon is linear. Dolan and McLaughlin (1967) find that the normalized shape of the decay curve for 575 laboratory sample measurements varies between the rather close limits shown in Fig.5.4. Data were analyzed in groups of 50 and these are compared on an equivalent basis. It is interesting to see from results at the periods used by Dolan and McLaughlin that it is possible to define a normalized standard decay curve with a fairly consistent logarithmic shape and a decay constant in the range 0.1—1.0 s. Hohmann et al. (1970) and other investigators have documented these characteristics in the frequency and phase domains.

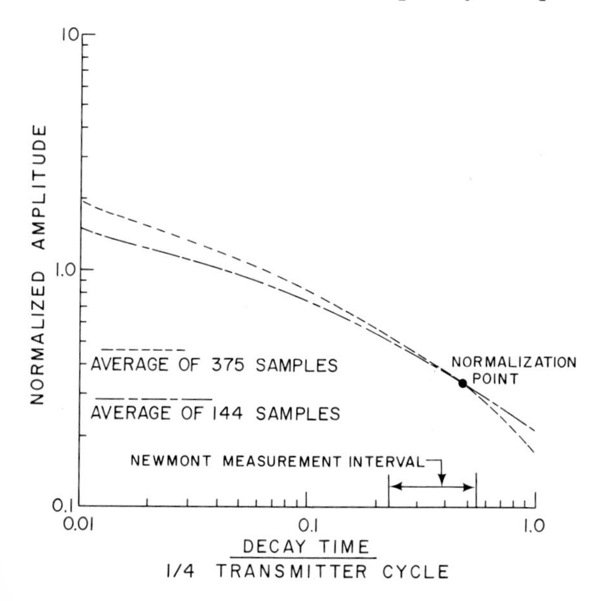

Fig.5.4. Normalized shape of a group of IP decay curves. After Dolan and McLaughlin (1967). A pulsed square wave of 2 s on positive, 2 s off, and 2 s on negative, etc. was used. The normalization point is arbitrarily chosen to give the best fit to the family of curves.

The observed fact that the time constant of the primary IP effect does not vary too widely at these periods is an important consideration in the design of an IP receiver.

It may be concluded from this that the amount of polarization is well expressed by the ratio of the amplitude of the decay curve to the primary voltage. Also, using the information from the standard decay curve (Fig.5.4), it is possible to optimize the time interval during which most IP measurements can be made, at least as far as obtaining first-order effects is concerned.

Noise characteristics of the ground

When current is not being put into the ground at the IP transmitter, the voltmeter-receiver will still measure the active electrical condition of the earth and the nearby environment, which may not be too quiet. These ambient voltage conditions are referred to as noise.

The wire that is laid out to measure the potential difference between the two electrode probes is essentially a grounded antenna which detects the IP and resistivity signals. This antenna receives noise voltages induced into the receiver electrode wires by local electrical transients from above the earth as well as those from natural oscillating earth currents. Therefore, it is sometimes helpful to shield the wire electrically or to bury it to reduce noise.

Noise can interfere with the desired electrical measurement and cause errors. Noise can be minimized by filtering or by overpowering the unwanted voltages, if the noise source is external to the measuring system. Noise that is internal to the system can usually be suppressed by the instrument designer or operator, if its presence is recognized and properly diagnosed.

Magnetotelluric or telluric noise is a rather complex topic that will be treated later in greater detail. Self-potential noise voltages are due to the presence of nearby electrochemical reactions in the earth. The instrument designer and field operator must be aware of the nature of these conditions that place a definite limit on instrument effectiveness at lower sensitivities.

Noise voltages are often random or incoherent in their frequencies and phase characteristics; that is, all frequencies are represented but bear no essential time relationship. These random noise voltages are additive as their squares:

$$V_{total} = \left|\sum_{1}^{k} V_i^2\right.$$

which tends to act in the measurer's favor.

It is sometimes possible to increase the signal-to-noise ratio (*S/N*) by repeating or integrating readings or by extending the reading time, in which case the *S/N* improves approximately as the number of readings.

The field operator can do much to suppress noise or even to eliminate it, especially when it is due to the local environment. Often some detective work is necessary, using methods of logic and deduction to identify the noise source by analyzing results from the measuring system. After trouble-shooting has found the local noise problem and isolated it, the elimination procedure may be incorporated in a regular field routine.

Signal-to-noise ratio, S/N, is useful in considering how difficult it will be to obtain signal information in the presence of noise. The ratio is sometimes given in decibels, where:

$$\frac{S}{N}\text{ (dB voltage)} = 20 \log \frac{\text{signal voltage } (V_s)}{\text{noise voltage } (V_n)}$$

Fortunately, even with a seemingly large amount of random noise, say a noise voltage of 5 and a signal voltage of 100, the increase in a receiver reading would actually be negligibly small.

If a sine-wave signal V_s (rms) is measured in the presence of a noise voltage V_n (rms), the indication ($\Delta V/V_s$) on an average responding voltmeter would increase by the following amounts:

V_s/V_n	$\Delta V/V_s$ (in percent)
∞	0.0
8	0.7
4	2.7
2	10.5
1	37.3

Ward and Rogers (1967) point out that a measured voltage signal V_s in itself usually consists of a sought-for message V_m plus an unknown amount of unwanted noise V_n. Although by this definition:

$$V_s = V_m + V_n \tag{72}$$

the addition may not be algebraic.

Noise can be reduced by the rejective filtering of unwanted frequencies or by increasing the power output of the signal source. However, a large power increase may be necessary to increase the voltage signal appreciably, as has been pointed out. In fact, to double the current in the same load, a fourfold increase in power is required. Extensive noise can often be eliminated with lower cost, weight, and power by rejection filtering at the receiver.

THE SIGNAL-GENERATING SYSTEM

Power requirements for the overall task to be performed by an IP field system are a primary design consideration. It is worthwhile to have the most suitable power source available for field surveying. Geologic requirements of the target region and physiographic conditions of the area will usually dictate the portability requirements of the power plant, electrical generator, and their interconnections. The power-to-weight ratio of a motor generator and its efficiency and dependability will certainly influence a designer's selection.

Motors and generators

IP power systems vary from simple, lightweight battery devices for shallow surveys to the large sources indicated in Table VI.

Battery systems are indeed ingenious, relying on sophisticated short-time measurements and rejection filtering at the voltmeter-receiver to conserve power. Two-cycle engines are light in weight but sometimes may not be too stable in output and require regular attention in the field in order to maintain proper performance. The power from a field vehicle can be used if an adequate voltage control device is installed. However, stable linkage from a power source (motor) to the electrical generator is not always easy to design in a simple, effective way. Power-line sources, when available, have been used for special surveys, for example, underground, when it is convenient to do so. Compressed air motors are often difficult to regulate precisely.

Four-cycle engines are a popular power source in IP field work because of their reliability and versatility. For higher power requirements, a three-phase generator has an advantage of being capable of producing a low-ripple current from a bridge rectifier, and its matching voltage step-up transformer is efficient on a power-per-pound basis.

IP generators are usually of the alternating-current variety so that the voltage can be more easily stepped up and regulated. There is a distinct advantage in using a higher frequency generator because comparable higher frequency step-up transformers are lighter in weight than lower frequency ones. Their weight is roughly inverse to the square root of frequency so that, by way of comparison, a conventional 60-cycle transformer is more than twice as heavy as a 400-cycle one:

$$2.6 = \sqrt{\frac{f_{400}}{f_{60}}} = \frac{W_{60}}{W_{400}} \tag{73}$$

Thus, for high-power, portable, electric power systems a higher frequency generator will save transformer weight, which is a major item in the overall

TABLE VI

IP power sources, giving general maximum output

Power source	General maximum output
Battery systems, including condenser discharge	50—200 W
Two-cycle engine	1.0—10 kW
Power takeoff from field vehicle	1—10 kW
Power-line source, when available	1—5 kW
Four-cycle engine	1—20 kW

weight of a portable IP transmitter. However, 400-cycle generators are becoming expensive and are in short supply, so that a generator of this frequency, while favored, may not be the generator of the future. Transformers can be damaged or destroyed by using frequencies below that for which they were designed. Also, lower frequency transients, for example, those created when a motor generator is allowed to die down, may cause harm. For this reason, an IP transmitter is matched to its power source and should be turned off before the power source is turned off.

Higher frequency noise created in the generating system as a result of switching and rectification may be transmitted with the inducing current. This unwanted effect is reduced and usually eliminated by electromagnetic attenuation and filtering at the voltmeter-receiver.

The IP transmitter

Once the optimum power source, energy transmission train, and generator are decided upon by the designer of an IP system, the details of the current sender (transmitter) must be specified. This unit is an automatic power-controlling system that may also regulate current to be put into the ground. It takes the raw electric power from the motor generator or battery and provides an optimum available current.

There are many different ways of sending current into the ground, starting with the manual switching of a simple dc source. Over the years, IP transmitters have progressively evolved from these basic approaches into fully automated systems. Figure 5.5 is a generalized block diagram of an IP transmitter that could be used for either frequency- or time-domain field work. Semiconductor switches and digital logic control devices are usually used whenever possible because of their high reliability and low weight and low power consumption. The wide range of temperature and humidity of the environment and rigorous field conditions call for considerable design engineering.

A current transmitter in frequency IP instrumentation must have current regulation to the same precision as that required by the received IP voltage signal. In older equipment, this was accomplished by using inductive-voltage and current-control devices employing the magnetic amplifier principle. More recently designed current senders take advantage of controlled rectifiers, which are silicon semiconductor switches with no moving parts. These can be put into a bridge rectifier type of circuit, as shown in Fig.5.6.

Current from the IP transmitter can be controlled to a constant value by using an output sensing resistor and by comparing its signal voltage level with that of a Zener reference diode. Any small difference in voltage between the current-sensing resistor and the reference diode and current-selector resistor is amplified and applied as a correction signal to a variable timing circuit. This voltage difference controls the delay time interval (or phase angle)

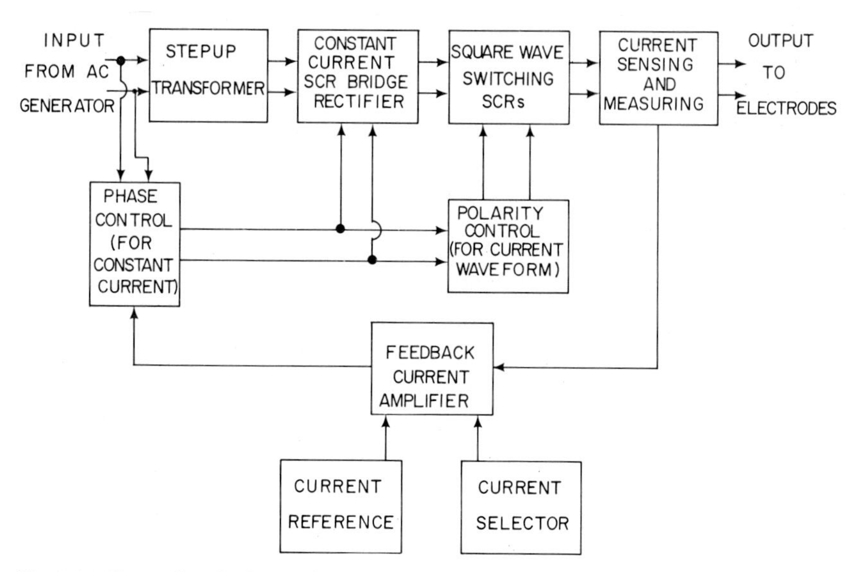

Fig.5.5. Generalized block diagram of a combination IP transmitter for both time and frequency domains.

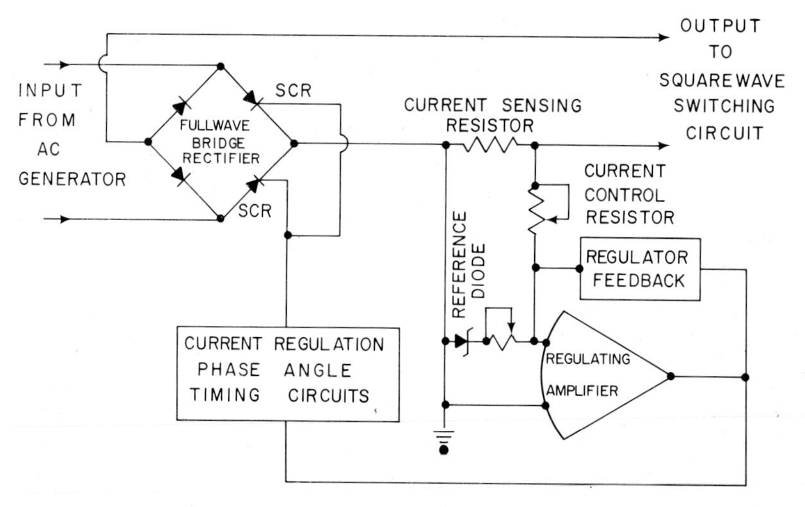

Fig.5.6. Full-wave silicon-controlled bridge rectifier for regulating a constant current. The general layout of the current-sensing and phase-control circuits is shown.

before triggering the input ac waveform to the silicon controlled rectifiers (SCR's) in the full-wave rectifying or ac bridge circuit. Precise current regulation is accomplished in delay timing or phase control by switching the SCR on later in the half cycle of the ac wave in order to reduce the output current to a predetermined level. Figure 5.7 shows the fraction of current

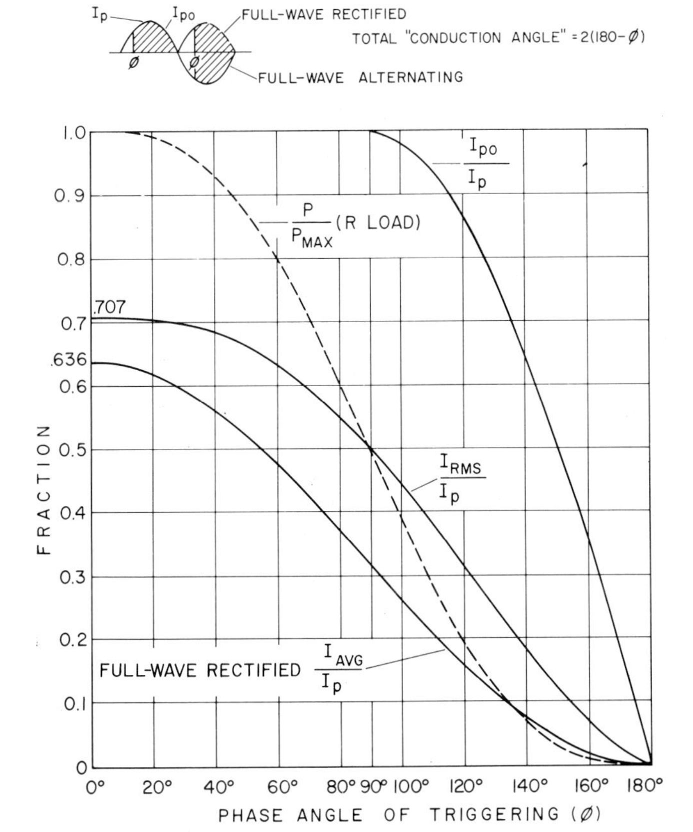

Fig.5.7. The fraction of available current plotted against phase angle of triggering of a silicon-controlled rectifier, illustrating the effective current control sensitivity and range of an SCR. I_{po} is the peak output current when the phase triggering angle ϕ is greater than 90 degrees.

that is transmitted as a function of conduction angle of the controlled rectifier. Note that at low phase angles the rate of change of current with change in phase angle is small, so that this control method is quite sensitive.

The phase-control method of current control makes it possible to synthesize practically any desired waveform by cyclically varying the conduction angle with time. In this way, a low-frequency sine wave can be produced by what is known as a cycloconverter triggering circuit.

Silicon controlled rectifiers are also used as current relay switches, chopping and reversing the rectified, controlled dc in the transmitter before it is put into the ground. A diagram of a simple square-wave switching circuit is illustrated in Fig.5.8. The frequency-domain current waveform is relatively easy to make, requiring only four SCR's in a bridge circuit, as shown. The time-domain waveform (Fig.5.9) requires perhaps six SCR's. Other

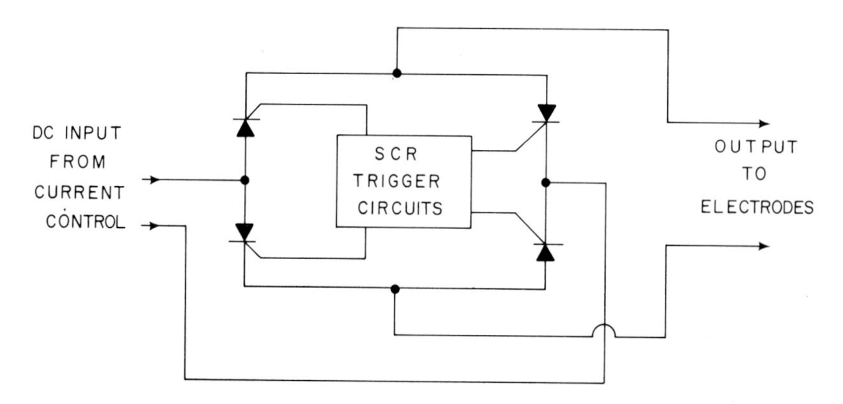

Fig.5.8. General schematic diagram of a square-wave chopper circuit using four SCR's.

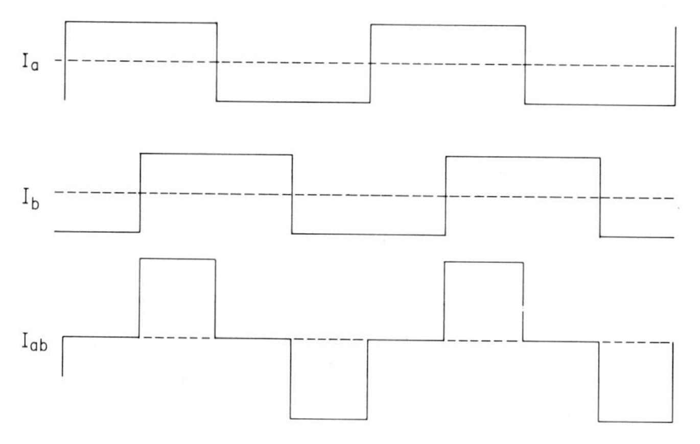

Fig.5.9. Two full square waves with a 90-degree phase difference, combined to form a pulsed square wave as used in time-domain IP survey work.

waveforms can be formed in a manner similar to phase control by multiple pulse-width control or by making selective harmonic reduction (Fig.5.10). The content of higher frequency harmonics can be suppressed or perhaps introduced for purposes of creating a simultaneous multiple-frequency waveform.

SCR's are noisy devices, but since the noise is much higher in frequency than the IP band of interest, this noise is reduced through the ground and is readily eliminated by rejection filtering at the receiver.

There is *always* the presence of danger when working with high-current and high-voltage circuits, which are numerous in IP transmitters. Extreme caution must be exercised, especially near the SCR's, if attempting to trouble shoot equipment or to operate the transmitter outside of its case. Safety measures will be discussed at greater length in the chapter on field methods.

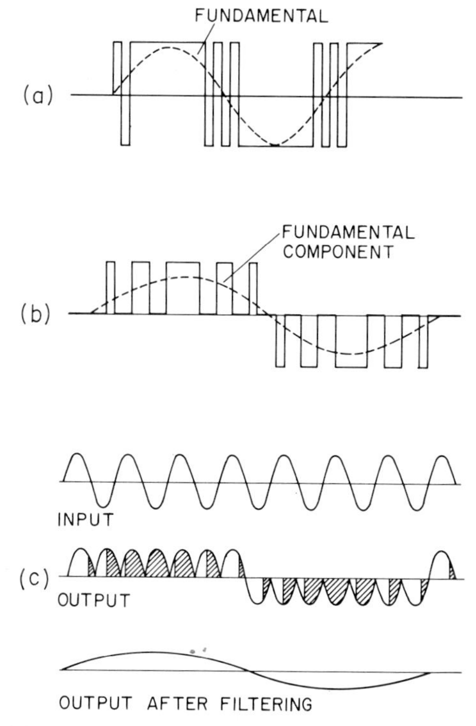

Fig.5.10. Current output waveforms from SCR control circuits: (a) a multiple pulse-width waveform, (b) harmonic reduction waveform, and (c) the cycloconverter waveform.

THE IP RECEIVER

Voltage measurement of the IP effect is made with an IP receiver. This basic electrical device ranges from a rather simple voltmeter or oscillograph to a completely miniaturized portable computer. An IP receiver is the nerve center of a field survey operation, and all of the really important survey results come from this instrument. It is usually portable and is read by the chief of the survey party. Each reading of the voltmeter-receiver is always important. The success or failure of the field program depends on the device, the manner in which it is read, and the interpretation of the values obtained from it.

Self-potential buckout

Both time-domain and frequency-domain IP receivers have used self-potential (SP) buckout controls, especially on older instruments. This control is sometimes automatic and has been eliminated on more recent IP receivers.

Self-potential is a dc or slowly varying electric earth potential due to telluric currents or a self-induced electrochemical reaction in a subsurface body. It seldom has a magnitude greater than 1.5 V but must be compensated for or 'bucked out' to within one or two millivolts where necessary. The control is basically a potentiometer connected across a supply voltage with the variable resistor arm connected to an input amplifier summing junction. The buckout system cancels any dc offset due to self-potential before the voltmeter input. It can also compensate for amplifier drift.

The time-domain receiver

The waveform of interest in the time IP method (Figs.1.1 and 3.13) can be detected and measured by a recording or integrating voltmeter. Indeed, this is the conventional manner of making the time-versus-voltage measurement, which can be done in several simple efficient ways.

The pulse IP receiver is a direct-coupled voltmeter that responds to direct current and extremely low frequency signals.

The first-order IP effect in the time domain, chargeability, is basically the ratio of IP response $\Delta\rho_t$ to resistivity ρ over a given period of time. This has been written $M = \Delta\rho_t/\rho$.

The physical determination of chargeability M itself should involve a simple quantitative aspect of the decay curve. This can be the ratio of secondary to primary voltage at one or more instants or centered intervals in time (t = 0, 0.1, 0.3, 1.0, etc.) on the decay curve, given as:

$$M_0 = \frac{V_{t_0}}{V_P} \text{ or } M_{0.1} = \frac{V_{t_{0.1}}}{V_P},\ M_{0.3} = \frac{V_{t_{0.3}}}{V_P} \tag{74}$$

By sampling the decay curve at geometrically spaced decay-voltage windows (eq. 74), one gains additional information on second-order IP changes, that is, the change in slope of the decay curve from a logarithmic form. Figure 5.11 is a schematic display of the decay curve intervals measured by the sophisticated Huntec Mark III time-domain IP receiver. In the field, the originally defined IP voltage ratio, V_s/V_p, is seldom measured immediately after cessation of charging current because of the errors that would be introduced from inductive electromagnetic coupling.

After the necessary time delay to minimize inductive EM coupling, the area A under the decay curve can be found by integrating the changing voltage with time over a given time interval:

$$A = \int_{t_1}^{t_2} V_t \, dt \tag{75}$$

Since the IP decay phenomenon is essentially linear with respect to the primary voltage V_p, to give the chargeability M, the area given in eq. 75 must be normalized by dividing by V_p:

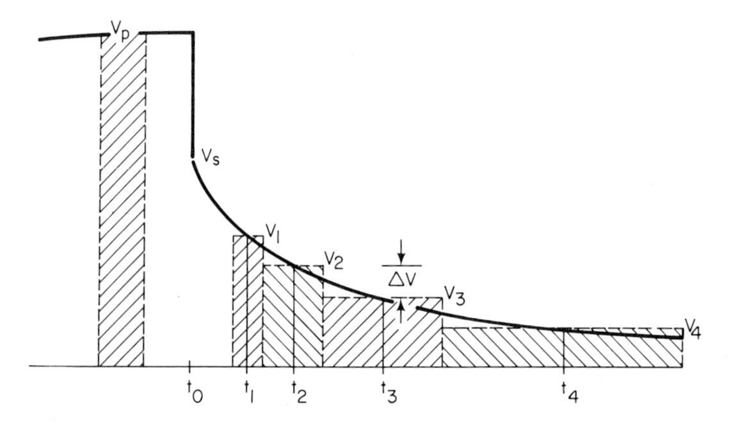

Fig.5.11. Decay curve intervals as measured with the Huntec Mark III time-domain receiver. A comparison of the areas, here based on a logarithmic decay rate, will give details on the shape of the decay curve.

$$M = A/V_p \tag{76}$$

Chargeability is generally normalized to M_{331} for a standard pulsed square wave of 3 s on, 3 s off, using a one-second integration time. This type of electrical determination is not too difficult to make, and electronic summing and integrating circuits can be designed into reliable field instruments. The measured IP voltage response from a voltmeter or oscillograph can be displayed in several ways, using digital values and normalized ratios of time intervals. For example, Mining Geophysical Surveys of Tucson plots field IP receiver values on a histogram, so that the noise patterns can be seen.

The Newmont receiver system is a popular time-domain receiver in North America, and its design has been discussed by Dolan and McLaughlin (1967). It features a self-potential cancellation circuit at the input, a self-triggering of a programmed cycle by sensing the primary voltage, a direct reading of chargeability M as the normalized area under the decay curve integrated from $t = 0.45$ to $t = 1.1$ s and a provision for reading the complement L of the chargeability as the normalized area over the decay curve. Figure 5.12 is a generalized schematic diagram of the Newmont type time-domain IP receiver. The Newmont receiver chargeability value is calculated to the M_{331} standard.

Measurement of the complement L of the chargeability is sometimes a practical quantitative way of obtaining information on the character or shape of the decay curve. The area above the decay curve is referenced to the zero level established by the decay voltage at the beginning of the measurement, which is 0.45 s after the induced current has shut off. As is shown in Fig.5.13 for the M_{331} standard, the ratio of areas L/M is set at slightly less

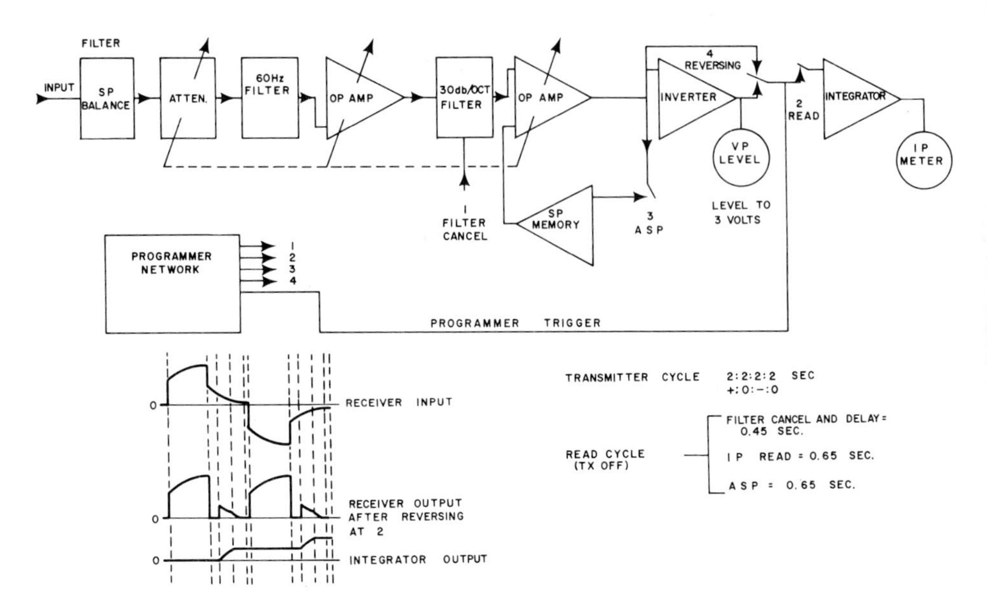

Fig.5.12. Block diagram of the Newmont-type time-domain IP receiver. After Dolan and McLaughlin (1967).

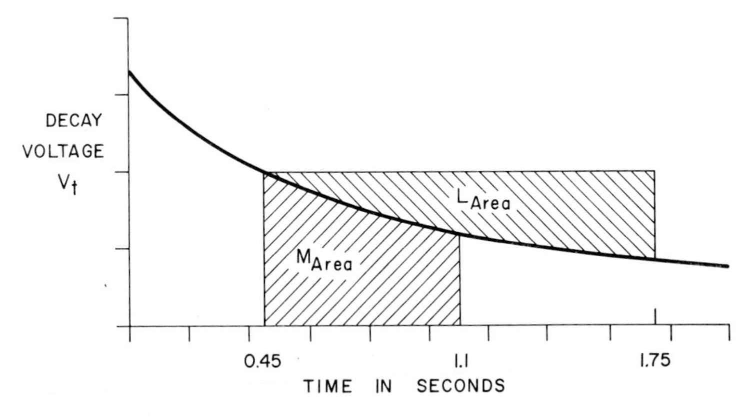

Fig.5.13. The IP decay waveform showing the areas L_{AREA} and M_{AREA} as measured by the Newmont-type IP receiver.

than unity and for standard decay curve of Fig.5.4:

$L/M \approx 0.8$

Departures of L/M from 0.8 indicate whether the decay curve is more or less steep than the standard and to a lesser extent imply the amount of curvature. This information can be useful for the interpretation of field data, if the meaning of variation of L/M is correctly understood (Swift, 1972). For example, variance of L/M can indicate the presence of electromagnetic coupling or 'cultural coupling' effects due to artificial conductors. Also, particle size in polarizing materials, composite decay curves, and geometric effects may be indicated.

Older time-domain IP transmitters and receivers were synchronized using electrical interconnections, and this dependence limited the types of electrode arrays that could be used. Newer pulse receivers can be remotely triggered by sensing the 'off' instant of transmitted voltage, so that the IP voltage can be detected during the transmitter off time.

The frequency IP receiver

The frequency-method waveform has been illustrated (Figs.1.2 and 3.11) in the way an unfiltered, noise-free polarization signal would be received at the input of a voltmeter. Usually voltages of two or more frequencies are measured separately from the current-controlled, full-wave transmitter. The input of most frequency instruments is ac coupled so that self-potential and potential electrode polarization effects are not measured. However, dc coupling is usually available as a selective input feature on the voltmeter-receiver so that very low frequencies and self-potential can be measured where necessary.

Scintrex of Canada has developed an interesting frequency-domain IP receiver which uses the information contained in the transmission of a single repetitive square wave. The frequency effect is electrically determined between the fundamental and the third harmonic components of the received waveform. In addition, the relative phase shift is determined between the fundamental and third harmonic absolute value components. This method does not require a radio link or synchronized crystal clocks.

McPhar has used a dual-frequency waveform and transmitted and measured two frequencies simultaneously. This method has the advantage of providing an immediate readout of frequency effect with one transmitted signal waveform allowing for rapid measurements with battery-powered equipment. This waveform is shown in Fig.5.14, which also summarizes other IP measurement methods.

A phase interlock system for voltage detection using frequency coincidence between transmitter and receiver has been used, and it is often available as an optional feature on frequency equipment.

The phase reference system can be:

(1) The input voltage waveform itself, the reference phase being set by received voltage zero crossings or a combination of rise time and peak voltages.

(2) Crystal-controlled clocks, present in synchronization in both the transmitter and receiver before the start of a survey.

(3) A telemeter link using radio signals for a phase reference.

(4) A direct electrical connection between transmitter and receiver for phase comparison purposes.

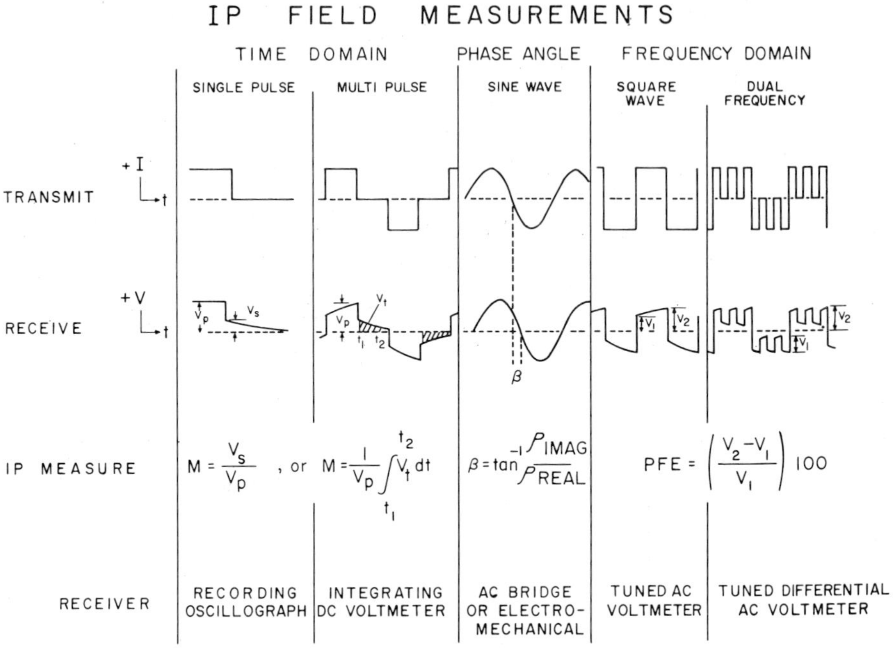

Fig.5.14. Summary of current and voltage waveforms and IP measuring methods. The conventional dual-frequency type waveforms are shown in Fig.1.2.

There is a considerable advantage in using a phase reference, or coherent, method of detection in the frequency IP technique if the necessary electronics are simple and reliable. Filtering advantages of the frequency IP method can be combined with the advantages of the decay waveform of the time method as described in the following paragraph.

Consider the charging current and voltage decay waveforms of Fig.5.15. If the voltmeter-receiver, which integrates the signal level in each interval, is successively switched from one time (or frequency) interval (t_1) to the next (t_2), a reasonable total decay curve or frequency spectrum can be derived directly from a series of field measurements.

By using a null-reading, difference-reading voltmeter or differential voltmeter circuits, frequency effect is often read directly on an expanded scale meter. The voltage reading from which resistivity calculations are made is taken from the nulling potentiometer dial.

The frequency IP voltage display meter often has a controllable damping factor. Most ac voltmeters are average reading instruments (Fig.5.16) calibrated in rms volts. Figure 1.2 shows a peak-to-peak voltage amplitude and Fig.3.11 shows a half-peak voltage amplitude. For a half-peak voltage of unity, the peak-to-peak value will be two, the rms voltage of a sinusoidal wave 0.707, and the average voltage 0.637.

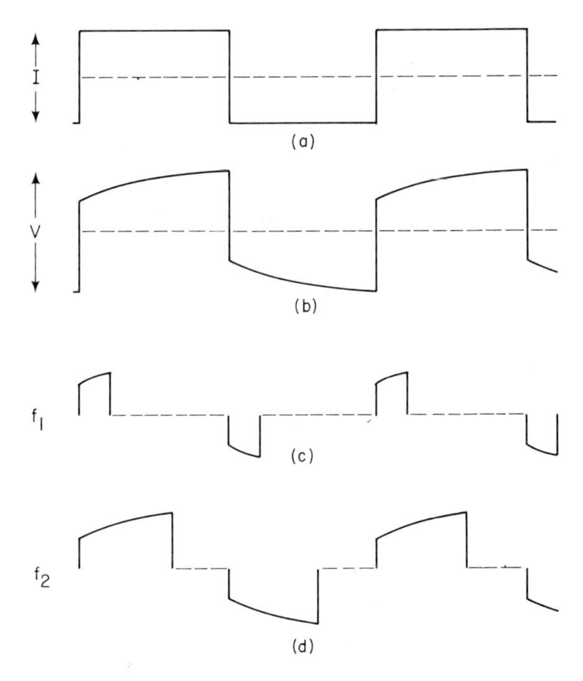

Fig.5.15. A method of synchronous, or coherent, detection showing (a) a single low-frequency constant current waveform, (b) the received voltage waveform, (c) a high-frequency portion, f_1, of the received waveform, and (d) a low-frequency portion, f_2, of the received voltage. The pulses shown in (c) and (d) can be electrically stored in the voltmeter-receiver and replayed, eliminating the off time to produce the waveforms of Fig.1.2.

Like the time-domain instrument, the frequency IP receiver is simply a specialized, sensitive, low-frequency voltmeter that has been made portable and rugged enough for field use. A simplified block diagram of a typical instrument is shown as Fig.5.17.

Digitized IP field equipment

K. L. Zonge and his associates in Tucson, Arizona, have designed a completely digitized field IP unit (Fig.5.18), which uses a digital computer to direct operations and analyze results. Zonge's IP transmitter sends out a series of square-wave current signals which are sampled by the digital computer and then compared in real time with the exact amplitude of the in-phase and out-of-phase voltage responses. The harmonic frequencies of the square waves are used in the analysis technique. The field data thus consist of a complete IP response spectrum with both real and imaginary components. The method is sensitive to the background polarization of unmineralized rocks and presents a 'signature' of the geologic units being

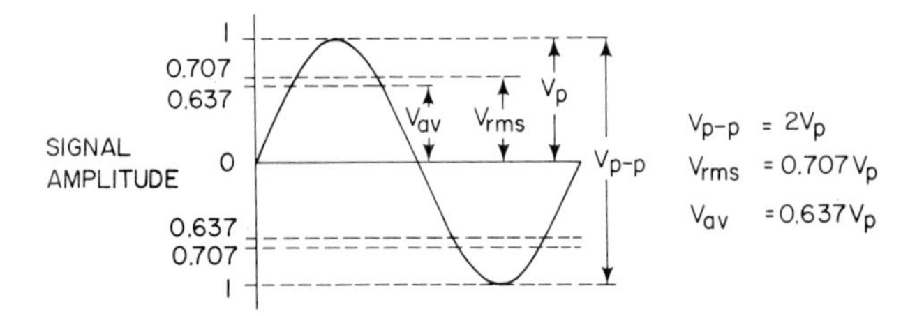

Fig.5.16. Half-wave (V_p), rms (V_{rms}), and average (V_{av}) voltages of a sinusoidal waveform.

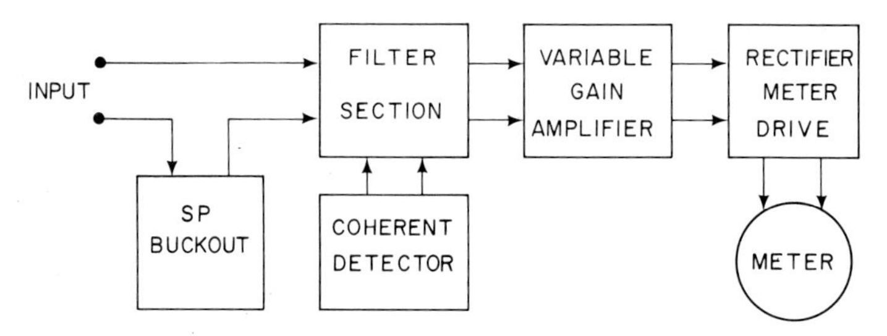

Fig.5.17. Block diagram of a typical frequency method IP voltmeter-receiver.

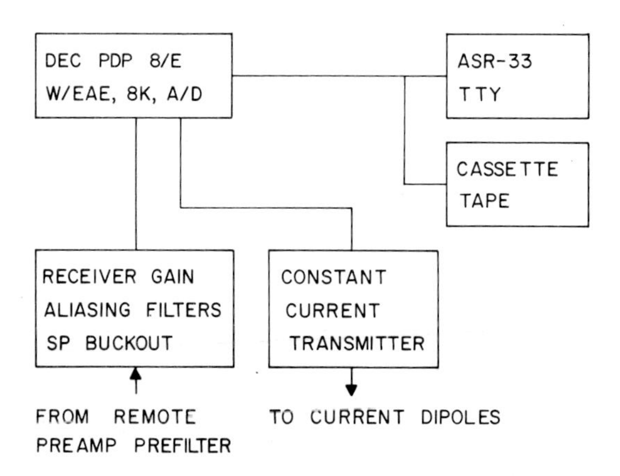

Fig.5.18. Scheme of a complex-resistivity field system showing main components.

traversed. Also, it appears that the various metallic constituents of mineralized rock make distinctive contributions to the polarization signal. It is also possible to make a separation between the electromagnetic coupling and the IP response and to analyze the effects of electromagnetic coupling. The electromagnetic response can also be analyzed to interpret the effects of depth and layering and to determine conductivity contrast, so as to supplement the resistivity interpretation of data. Furthermore, the electromagnetic response can be separated into a purely inductive component and a component due to EM wave reflections.

Design considerations from frequency analysis of a square wave

At first impression it may seem peculiar that a full square wave can be completely synthesized from a series of purely sinusoidal waves. As a matter of fact, any cyclically repeating waveform can be generated in this manner. This is not only a possible way of creating an arbitrary wave shape, but it is also a powerful technique for analyzing any given waveform to determine the harmonic content. It is this latter use of waveform study that makes Fourier analysis useful in finding the harmonic relationships between a sine wave and a square wave.

In order to perform this analysis it is necessary that the waveform be described by a Fourier series of the form:

$$f(x) = \frac{a_0}{2} + \sum_{n+1} (a_n \cos nx + b_n \sin nx) \tag{77}$$

where the coefficients of a_n and b_n are given by:

$$a_n = \frac{1}{\pi} \int_{-\pi}^{\pi} f(x) \cos nx \, dx \text{ and } b_n = \frac{1}{\pi} \int_{-\pi}^{\pi} f(x) \sin nx \, dx$$

For a full square wave with a dc component in the interval between $-\pi$ and π, as shown between the dashed lines of Fig.5.19:

$f(x) = 0 \quad \text{for} -\pi < x < 0$

and:

$f(x) = \pi \quad \text{for } 0 < x < \pi$

Then:

$$a_0 = \frac{1}{\pi}\left(\int_{-\pi}^{\pi} 0 \, dx + \int_{0}^{\pi} \pi \, dx \right) = \pi$$

$$a_n = \frac{1}{\pi} \int_{0}^{\pi} \pi \cos nx \, dx = 0$$

and:

$$b_n = \frac{1}{\pi} \int_{0}^{\pi} \pi \sin nx \, dx = \frac{1}{n}(1 - \cos n\pi) = 0 \text{ for } n \text{ even}$$
$$= 2/n \text{ for } n \text{ odd}$$

Therefore, the Fourier series for this square wave is:

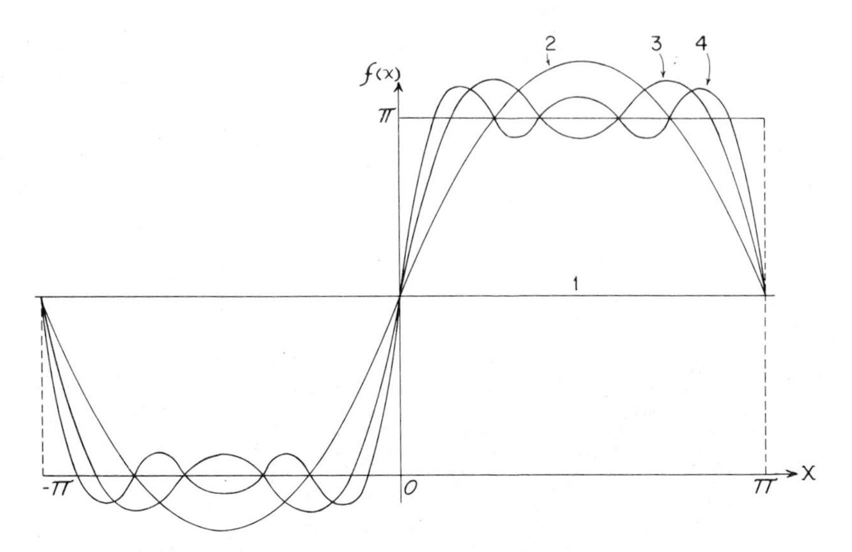

Fig.5.19. Fourier synthesis of a square wave with a dc offset.

$$f(x) = y = \frac{\pi}{2} + \frac{2\sin x}{1} + \frac{2\sin 3x}{3} + \frac{2\sin 5x}{5} + \ldots \tag{78}$$

The first $\pi/2$ is the dc term, labeled 1 on Fig.5.19. It can be set equal to zero for a full square wave, symmetric about the x-axis. The second term, $(2 \sin x)/1$, is a pure sine wave, which is the fundamental frequency of the square wave. The successive sums adding the next two higher ordered terms are labeled *3* and *4* on the diagram. After adding several orders of terms, the square wave becomes fairly well formed, as is seen on Fig.5.20. For this particular square wave, as used in phase and frequency IP methods, the following comments are applicable:

(1) Most of the signal content is in the first harmonic, whose frequency is equal to that of the square wave. Tight filtering would allow only the first harmonic to be passed. Note that the peak amplitude of the first harmonic is greater than that of the square wave, a fact which allows a greater peak voltage after rejective filtering of the higher frequencies of the square wave.

(2) Only odd harmonics are present, that is, $n = 1, 3, 5, 7$, etc.

(3) The individual amplitude of a harmonic falls off as $1/n$ where n is the order of the harmonic.

(4) The total effect (the power) of an individual harmonic on the form of the square wave falls off as $1/n^2$. This means that most of the energy of the square wave is contained in the fundamental harmonic. Figure 5.21 is a power spectrum of a square wave.

(5) There are higher amplitude oscillations near the corners of the square wave, caused by wave interference from higher harmonics.

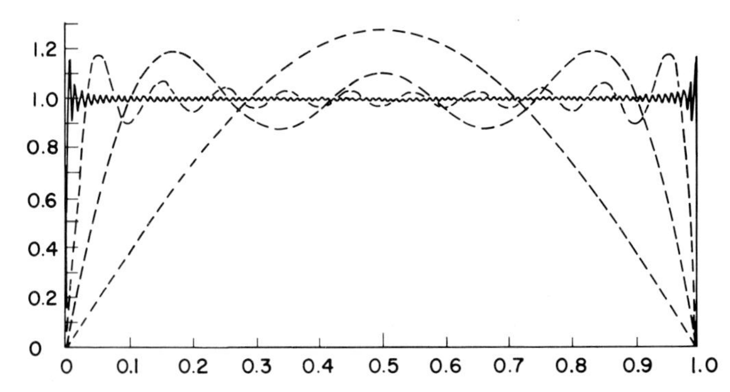

Fig.5.20. A Fourier series showing the first term, the sum of the first three terms, and the sum of the first 100 terms. Note the transient peaks at the ends of the higher order waveforms.

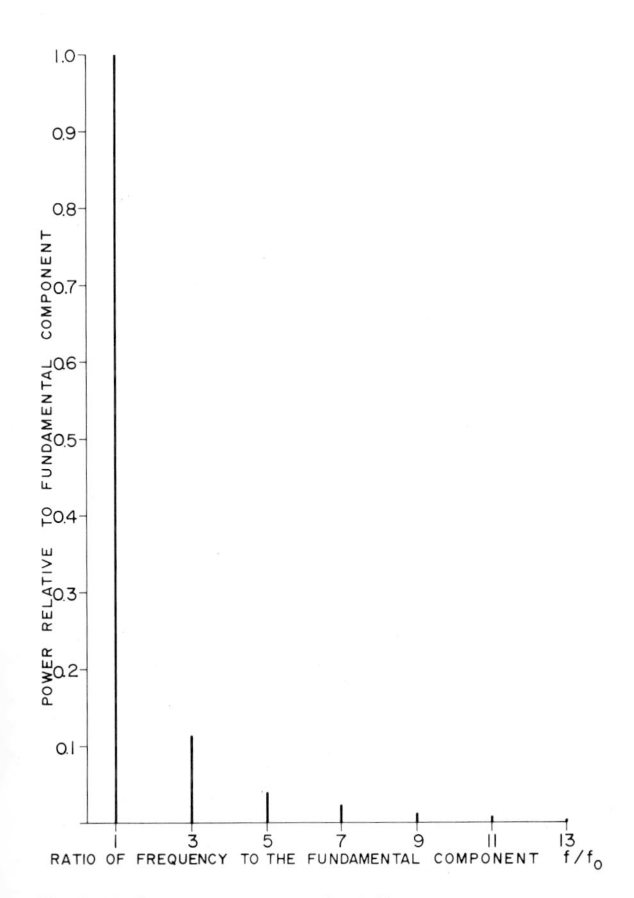

Fig.5.21. Power spectrum of a full square wave.

Filtering

The use of electronic rejection filtering to reduce or to eliminate unwanted noise is a sophisticated way of improving the signal-to-noise ratio of the IP receiver. The advantages over increasing transmitter power are that filters are reliable, light weight, require only low power, and are relatively low in cost.

Several different kinds of electronic noise rejection filters are normally used in an IP receiver. The subject of filter design is rather extensive and beyond the scope of our immediate concern; however, some discussion is necessary. Filters are usually inserted in the receiver design only to overcome some particular noise conditions, because their presence can cause problems of signal loss, power consumption, phase problems, and system maintenance considerations. Filters are either passive or active, and with the advantages of miniaturized solid-state circuitry there is a tendency for more active devices, which still consume only a minor amount of power. A good treatment on filter design is to be found in *A Handbook on Electrical Filters* by White Electromagnetics, Inc. (1963).

Some filtering is usually done toward the input or 'front end' of the receiver, before preamplification, if possible. For example, there often are no IP frequencies of interest above 10 Hz, so that low-pass filtering to reject power-line frequencies, radio station interference, and sferics is appropriate. Operational amplifiers are very convenient devices in IP receivers because of their many desirable characteristics, and it is a simple matter to provide a filtered feedback loop around them. Feedback R-C transistor network filters in parallel or twin-tee circuits provide good bandpass or notch characteristics where needed. Low-pass or bandpass filters are of the Butterworth (maximally flat) or Tchebyscheff (equal ripple) types; the Butterworth has a lower insertion loss and the Tchebyscheff a greater filtering efficiency. Inductive filters are seldom used in IP receivers because of their size, weight, and oscillatory 'ringing' characteristics.

Although filter design is initially approached from an idealistic or theoretical standpoint, electronic components do not behave ideally; that is, a resistor has some capacitance and a capacitor some resistance, so that some experimentation is necessary for proper filter design.

Besides the conventional analog waveform filters of older electronic equipment, the modern designer also has at his disposal several numerical or digital filtering techniques. Data taken numerically in the time domain can be filtered numerically by summing weighted samples at a series of successive time increments. Digital filtering can be exactly equivalent to analog filtering.

Coherent detection is an electronic filtering method basically using the synchronous detection method mentioned earlier in which the voltmeter-receiver compares an internally generated phase and voltage reference with the desired external signal. Noise of a different phase or frequency is strongly suppressed.

Chapter 6

TELLURIC NOISE AND ELECTROMAGNETIC COUPLING

Induced-polarization surveys in the field are limited by two principal external factors that restrict the sensitivity and usefulness of the upper and lower measurements in the time and frequency domains. The first factor is telluric noise and the other is electromagnetic (EM) coupling. Telluric noise is generally more of a problem with direct-coupled dc voltmeters, such as those used as IP time-domain receivers, whereas electromagnetic coupling is more bothersome to frequency-domain systems. This chapter will discuss the pertinent facts of these phenomena from the standpoint of their effects on IP surveying.

TELLURIC (EARTH) CURRENTS

Small, slowly varying electric currents continually flow in the earth. Although the direction of a current can vary, it is generally parallel to the earth's surface. At shallow depths, there are high-frequency components that diminish with depth. The quasi-random telluric currents that are induced to flow in the earth are principally caused by small variations in the earth's magnetic field that have their origin external to the solid earth. Electromagnetic waves are propagated downward from space to the solid earth, and the lower frequency energy penetrates to considerable depths. The depth to which electromagnetic energy penetrates the earth is referred to as the 'skin depth', the depth at which the field strength falls to $1/e$ of the value at the surface, which is a function of the frequency of the energy and the resistivity of the earth.

From the laws of propagation of planar electric and magnetic fields on the surface of an electrically homogeneous earth, skin depth δ is defined as:

$$\delta = \sqrt{\frac{2\rho}{\omega\mu}} \tag{79}$$

which can also be written $\delta = (1/2\pi)\sqrt{\rho T}$. The earth's magnetic permeability μ is usually very low, about that of free space. In terms of commonly used units, magnetic field H is measured in gammas, electric field E in millivolts per kilometer, and resistivity ρ in ohm-meters, yielding:

$$\delta = \frac{1}{2\pi}\sqrt{10\,\rho T}$$

or, after Caigniard (1953):

$$\rho = 0.2T\left(\frac{E}{H}\right)^2 \tag{80}$$

where T is the wave period in seconds and the ratio, E/H, is the 'wave impedance'.

There is a wide distribution of frequencies of electromagnetic radiation impinging on the earth's surface, but on the average, low-frequency energy makes up the greater proportion of its content. Figure 6.1 shows a typical power spectrum of the telluric field for Japan and for New England, showing the low-frequency energy distribution.

Telluric noise often has a maximum horizontal direction, as shown in Fig.6.2. This direction is usually influenced by variations in local subsurface resistivity, such as may be caused by faults or veins, although there can also be a preferential direction (polarization) of the incoming waves. The proximity of a large conductive body, such as the sea, influences the predominant direction of telluric currents. Knowing the direction of polarization in the ground, if any, one may be able to minimize telluric noise by laying out traverse lines at right angles to the direction of maximum telluric voltage amplitude.

Sometimes the amplitude of natural telluric fluctuations is relatively low, and such conditions are referred to as 'quiet' as opposed to 'disturbed' field conditions. Quiet or disturbed conditions occur over very wide areas. The intensities of magnetic disturbances and associated telluric currents increase from low to high latitudes up to about 65° magnetic latitude in the auroral zones.

Magnetic storms commence at almost the same instant all over the earth. The character of individual storms varies, but following this sudden commencement, the magnetic field increases during the initial phase (Fig.6.3) and then may vary for an hour or so, followed by a gradual recovery, which may take as much as several days. The natural magnetic field variations with periods of 0.1 s to 10 min. that are sometimes associated with magnetic storms are called micropulsations, and these have been classified into several types according to frequency. A portion of the spectrum shown on Fig.6.4 shows the micropulsations.

The magnetosphere

Weak as it is, the earth's distant magnetic field fends off the steady stream of hot charged particles emitted from the sun. The solar wind travels along solar magnetic flux lines and consists of an ionized plasma, which also transmits energy by wave motion. Particle and field studies conducted from the space probes have been spectacularly successful in measuring the solar-wind parameters. High-velocity plasma particles are deflected at a distance from earth of about 10 earth radii at the magnetopause, as shown on Fig.6.5. The earth's magnetosphere, a hollow within the solar-wind environment, contains

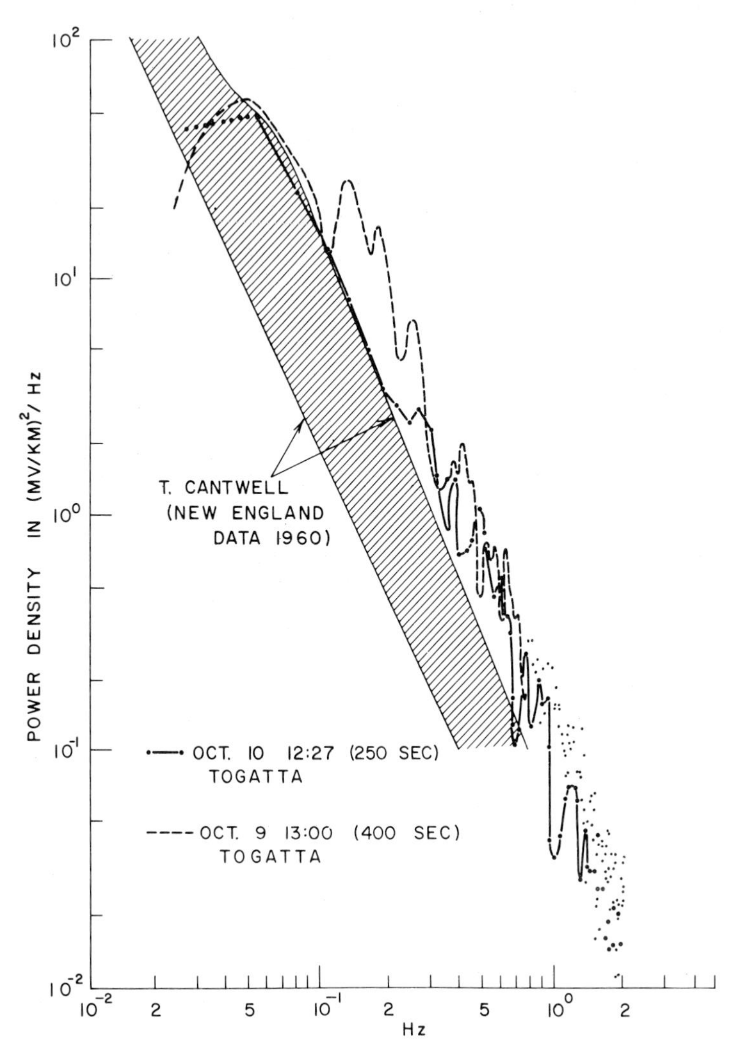

Fig. 6.1. Power spectrum of the telluric field in Japan compared with fields in New England. After Kunori et al. (1969).

within its bounds the earth's magnetic field, the radiation belts, and the earth itself. Solar plasma enters the magnetosphere along the dayside polar cusps through which it impinges on the earth's polar cap regions to form areas of particle activity known as the auroral ovals. Plasma consisting of ions moving in a complex fashion with respect to the earth's magnetic field rather than under solar influence also exists within the magnetosphere.

The solar wind has an ion-particle velocity of 300—500 km/s, and the normal wind contains about 10 protons and electrons per cubic centimeter. The solar wind is not constant in velocity or particle content, and solar-wind parameters can increase considerably during periods of solar activity and

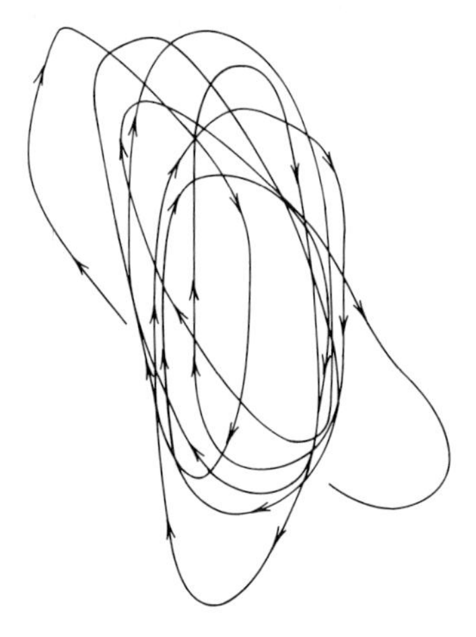

Fig.6.2. An electric field hodograph showing polarization of telluric noise extending over seven periods in the 1.3-Hz band. A smooth variation in orientation can be seen in which the ellipse axis rotates 20 degrees clockwise in an almost meridional direction. After Vladimirov and Kleimenova (1962).

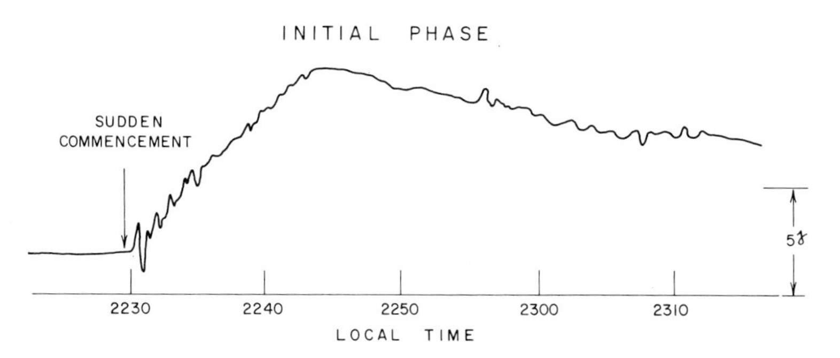

Fig.6.3. A recording of a magnetic field disturbance showing the initial phase of a bay with superimposed micropulsations. Not all storms have a sudden commencement; those that do have been called 'bps' for 'bay with pulsations and sudden commencement'. After Ward (1963).

gusty conditions can exist. The means of energy propagation is similar to that for acoustical energy in solid materials; energy is transmitted within a plasma by two modes: by transverse hydrodynamic oscillations, called Alfvén waves, or by a longitudinal hydrodynamic wave motion. Solar-plasma ions are turned aside by the earth's magnetic field in a collisionless shock front at the magnetopause. This creates an interface between the solar plasma and the earth's magnetic field along which hydromagnetic waves can be propagated, if such wave motion is started by a solar-wind disturbance.

The magnetosphere behaves like a huge bag containing energetic particles and plasma and swells and contracts under the influence of the solar wind.

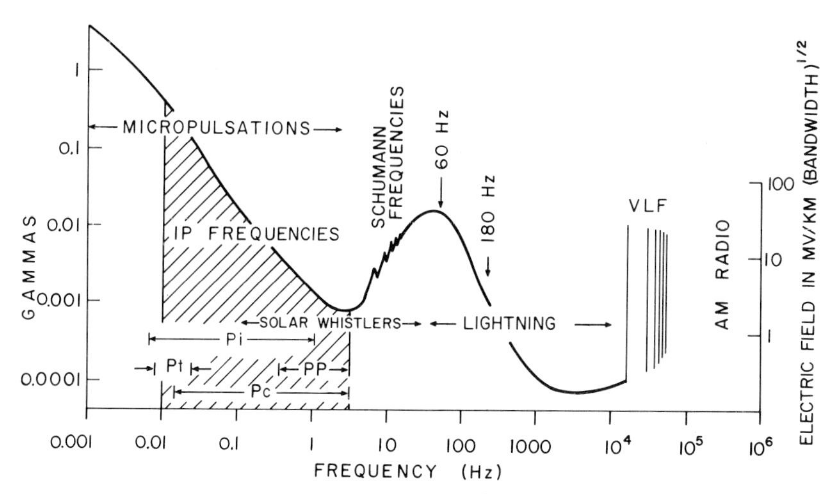

Fig. 6.4. The low-frequency electromagnetic field spectrum.

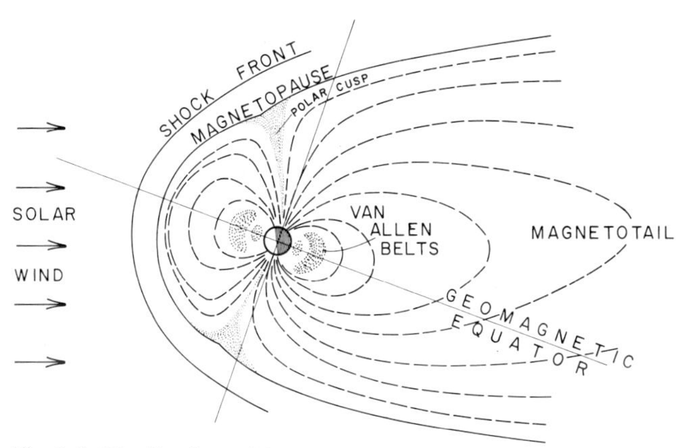

Fig. 6.5. Idealized earth's magnetosphere.

The sunward side is squashed by the solar wind, and lines of force are combed downwind into a long stretched-out tail. The magnetosphere oscillates like a bowl of jelly with characteristic time constants and periods, depending on its size. The recurrence of perturbations, called 'magnetospheric substorms', suggests that the geomagnetic cavity stores up energy until it is released by a triggering instability.

Micropulsations

The ion wave motion in the solar-wind plasma and in the magnetosphere generates a low-frequency electromagnetic radiation. This low-frequency energy is received at the earth's surface as micropulsations, observed as small magnetic field variations or as small electric earth currents. These small electrical pulses, called magnetotelluric currents or voltages, cause trouble to the voltage-detecting systems for IP field measurements.

Recent knowledge of micropulsations is summarized by Jacobs (1970) and in reviews by Saito (1969) and Troitskaya and Gul'elmi (1967). The topic is discussed regularly in articles in the *Journal of Geophysical Research*, a periodical of the American Geophysical Union. Knowledge of this and associated phenomena is continually being expanded. Because of the complex problems associated with understanding the physics of the upper atmosphere and the interactions of extraterrestrial plasma ions and because of instrumentation difficulties in measuring and interpreting the small magnetotelluric fluctuations, there is still much to be learned about micropulsations. The more continuous type micropulsations have sinusoidal amplitude changes; other micropulsations are rather irregular in form.

Micropulsations can be classified according to:

(1) Their recorded character, such as period, amplitude, and time of occurrence.

(2) Correlation with other phenomena, such as magnetic storms, aurora, and very low frequency (VLF) emissions.

(3) Origin, based on the mechanism of generation.

Although origin (3) or correlation (2) would be better classification systems, these matters are not clearly known or understood, thus the less satisfactory morphologic descriptions of recorded character must be used. Originally, micropulsations were classified as continuous pulsations (Pc), pulsation trains (Pt), or giant pulsations (Pg). In 1964, a new system was adopted, based on period, with Pc 1 consisting of continuous pulsations in the period range 0.2—5.0 s and Pc 2 in the period range 5.0—10.0 s. Irregular pulsations in the period range 1.0—40 s are referred to as Pi 1. Irregular pulsations are sometimes called W-events because of the similarity of the recorder trace to the letter 'W'. In IP field surveying, the only frequencies of concern are those that are characteristic of the Pc 1, Pc 2, and Pi 1 micropulsation types, although for long recording time, the Pc 3 type (periods 10—45 s) becomes important.

Pc 1-type micropulsations seem to be caused to some extent by ionized particles travelling in bunches within the magnetosphere along the earth's magnetic flux lines, oscillating between conjugate points at higher geomagnetic latitudes, as shown on Fig.6.6. Ion oscillations at cyclotron frequencies occur in the magnetospheric cavity and these wave amplitudes are reinforced at resonant mode frequencies, producing longer period micropulsations. A record of Pc 1-type continuous pulsations is shown as Fig.6.7, where an expanded portion of a frequency-against-time record illustrates frequency composition. Figure 6.8 shows a power spectral analysis of the record shown in Fig.6.7.

There are several other recognized types of continuous (Pc) and irregular

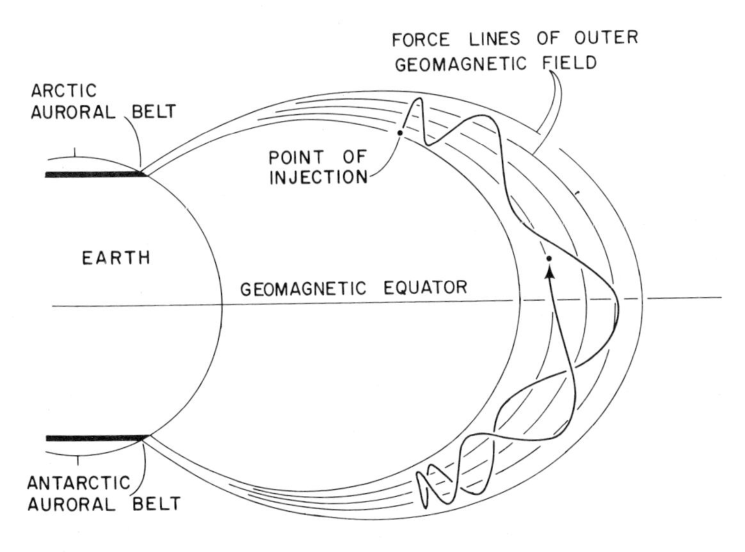

Fig. 6.6. Channeling of geomagnetically trapped particles in the earth's magnetic field, showing trajectories of injected particles.

(Pi) micropulsations, such as pearls (PP) at higher geomagnetic latitudes, consisting of beating frequencies or repeated bursts. There also are bursts and changes of frequency with amplitude in repetitive patterns. Pi's tend to be nighttime phenomena and Pc's daytime phenomena. To the exploration geophysicist, voltages caused by these micropulsations that tend to mask IP measurements are indeed 'noise', but to the academic geophysicist studying the solar wind and the characteristics of its interactions with the magnetosphere these voltages are meaningful signals.

Man-made electrical noise

Fortunately, there are few man-made electrical devices that operate in the IP frequency band of interest. Power-line noise and higher frequency interference can be filtered at the IP voltmeter-receiver. There usually are direct-current electrical devices, such as hoists, electric railways, and dc power lines, operating near mines and cultured areas that cause some voltage interference. When dc-powered systems are turned on or off, large transient low-frequency magnetic fields are created. Leakage currents and transient voltages due to these dc sources have often been known to affect IP readings adversely. Sometimes it is necessary to carry out IP surveys in noisy areas at night or on weekends when the noise is much less.

In areas of exploration activity, there may be several IP crews present and the signals generated by one crew can interfere with voltages being read at another crew's receiver. A mutually agreeable IP survey schedule may be necessary to eliminate interference so that all can obtain better data.

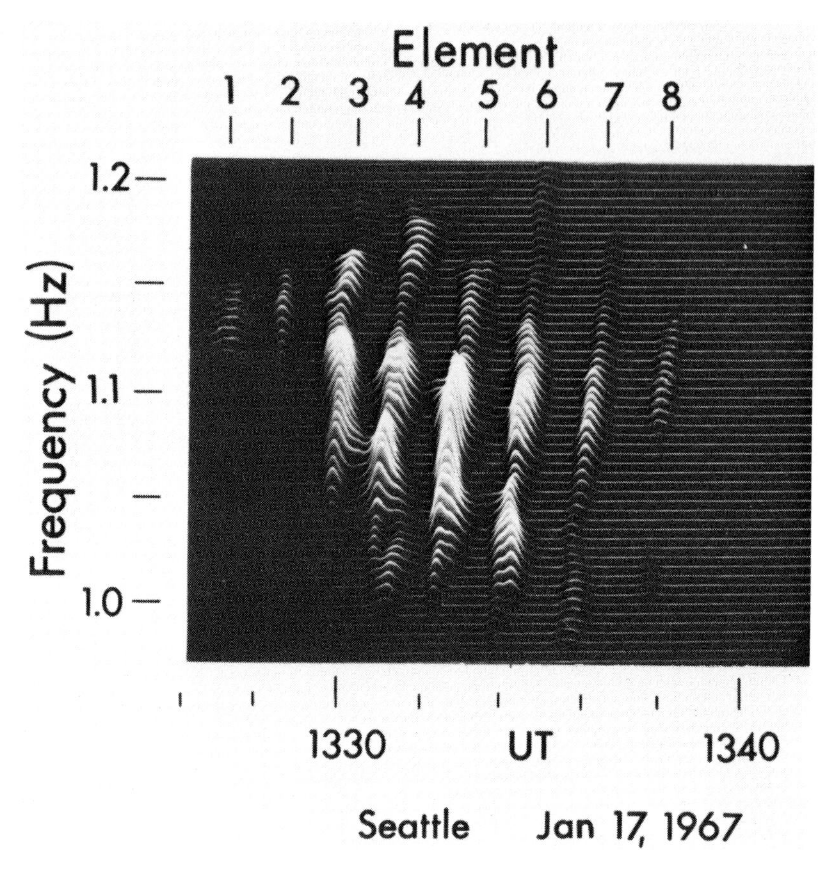

Fig.6.7. A frequency-against-time record (sonogram) of a Pc 1 micropulsation burst. After Kenney et al. (1968).

Sometimes leakage currents and low-frequency noise voltages existing near IP power-generating systems are large enough to be received by a nearby voltmeter-receiver. Shielding the power source, the receiver, or the receiver wire and increasing the distance between transmitting and receiving units are the usual methods of minimizing these noise effects. While shielding of a potential line may give rise to a capacitive coupling with the shield, this type of an effect is relatively constant and thus can be compensated.

Surface wind can cause significant noise by oscillating the receiver wire in the earth's magnetic field if the wire is suspended above the earth. Such vibration creates a 'wind noise' that can be minimized by fixing the wire to the ground or even burying it.

Fortunately, the amount of natural telluric noise is at a minimum in the region near one cycle per second, which is also the order of the decay constant of the IP effect. This is illustrated in a general fashion on Fig.6.4.

In the region near 8 Hz, there is a natural energy peak of the basal

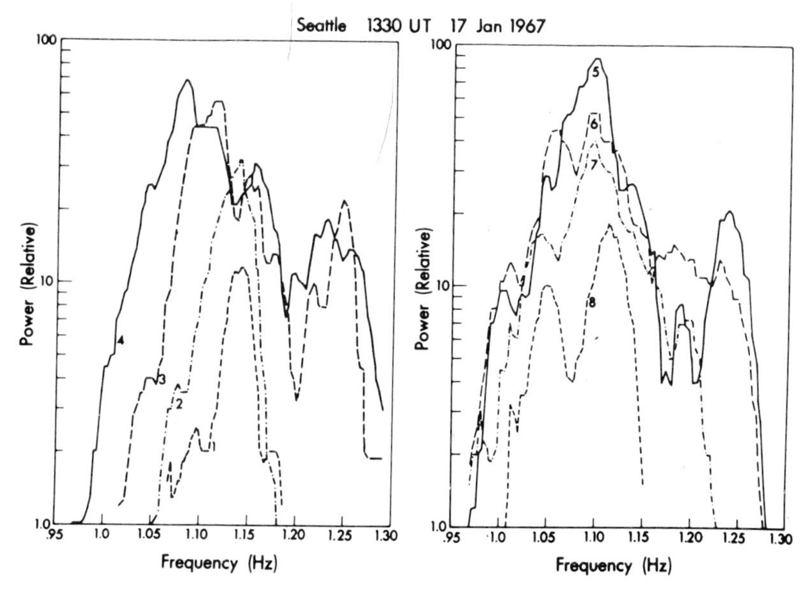

Fig.6.8. A power spectral analysis of the Pc 1 micropulsation record shown in Fig.6.7. After Kenney et al. (1968).

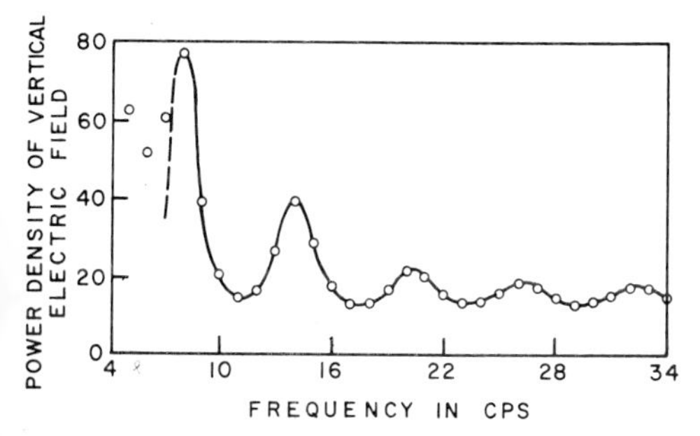

Fig.6.9. Resonance peaks of Schumann cavity resonant modes. After Madden and Thompson (1965).

resonant mode of electromagnetic radiation reflected between the earth and the ionosphere, as was first pointed out by Schumann (1952). Electromagnetic energy at this and somewhat higher frequencies is effectively trapped within the earth-ionospheric cavity. This entrapment is due to the dimensions of the cavity and to the higher conductivity of the bounding spherical reflecting surfaces. Figure 6.9 shows the relative amplitude of the resonant peaks. The energy source of the Schumann resonance phenomenon is probably equatorial thunderstorms. There also exist diurnal variations and directional dependences to this kind of radiation. The Schumann resonance effects are not as pronounced in their voltage amplitude as the much lower frequency telluric noise and are more distant in frequency than the commonly used IP frequencies.

Sferics

Atmospheric electrical phenomena, known as sferics, consist mainly of pulses of electromagnetic radiation due to lightning strokes. Most of the energetic content of these bursts is much higher in frequency than those in the IP region of interest, but sferic effects can sometimes be bothersome. As a generalization, if IP data are being affected by sferics, a thunderstorm is uncomfortably close. It becomes dangerous to personnel and equipment to carry out electrical geophysical surveys too near a thunderstorm.

Methods of avoiding magnetotelluric noise

Micropulsation activity is greater during periods of sunspot maxima, and noise from this source can be predicted from sunspot correlations. It may be practical to install semipermanent potential electrodes in an area to monitor the telluric activity with a voltmeter and chart recorder.

If there is a choice in making IP voltage measurements over low or high-resistivity ground by exchanging positions of the receiver with the transmitter, telluric noise effects will generally be minimized by reading the IP voltage on the lower resistivity ground. Telluric currents are relatively constant through adjacent high and low-resistivity areas (see eq. 22), so received telluric voltages will be larger over the high-resistivity material.

Telluric noise is generally greater in areas of higher ground resistivity (eq. 75) than in areas of lower ground resistivity. However, the received signal voltage can be more readily increased in higher resistivity regions, so that signal-to-noise ratio is lower in low-resistivity regions, as will be discussed in more detail later.

Band-pass frequency filtering is very effective in eliminating noise from most ac coupled IP voltmeter-receivers. On other receivers, digital filtering is sometimes done, using statistical or numerical techniques. The area under a decay curve can be made more accurate by the simple process of averaging the areas under a number of decay curves. If the average random error of the value for the area under a curve is D, then the error E in N area determinations is $E = D/N$.

As a field test of the amount of telluric noise, one can simply note successive receiver voltage readings and observe the average deviation of values. If telluric noise becomes a serious problem on a particular set of voltmeter-receiver readings, try reading the station at other times of day, such as 2, 4, 8, or 12 hours before or after the noise-problem time. The magnitude of telluric voltages changes with time during a 24-hour period and a better time can be found, perhaps at night. Also try changing the azimuth at which readings are taken, because of possible directional polarization, as shown on Fig.6.2. Reversal of the voltage and current electrode pair may show that for some reason one electrode pair is much noisier than the other. Noise effects are often greater at higher potential electrode resistances, thus the electrical contacts should be improved if these conditions are suspected. As a 'brute force' approach, it is always possible to improve the received signal level by increasing the current to the transmitting electrode pair.

COUPLING OF ELECTRICAL CIRCUITS

As used in the remainder of this chapter, the term 'coupling' will refer to the electrical linkage relationship between an IP transmitter and a receiver. Although the primary interest will be in electromagnetic or inductive coupling, two other forms of coupling exist: resistive and capacitive coupling.

Resistive coupling

As was discussed in Chapter 2, resistive coupling is the means by which the resistivity of the ground is measured. The current transmitter and the voltage receiver are electrically coupled together through the intervening earth, and the interpreted amount of resistive coupling measured with respect to transmitter and receiver locations determines the resistivity of the earth and the shape and location of any resistivity contrasts in the vicinity. Resistive coupling cannot be ignored when discussing and treating electrical coupling in its entirety.

Capacitive coupling

It can be said that capacitive coupling — as the term is used in IP surveying — is a voltage change due to electrical leakage or displacement currents that causes errors in IP measurements. Since induced polarization is a capacitance-like phenomenon, capacitive coupling can be a serious problem in IP work.

A common occurrence of measurable capacitive coupling arises where wires leading to transmitter and receiver electrodes are encased in a common insulating sheath. Electrically, this is a two-dimensional capacitor, the capacitance of which is usually given by the manufacturer or can be measured on an impedance bridge. In field surveying situations, measurement of wire capacitance effects is not easily done, but the effects can be calculated, as reviewed in the following paragraphs.

To find the capacitance between wires, let a fixed potential difference be applied between conductors so that the charge per unit length on each conductor is Q_L. The surface of the wire is an equipotential surface, and therefore an equipotential circle (Fig.6.10) will coincide with the wire surface. The heavy circles of radius r on Fig.6.10 and the center-to-center

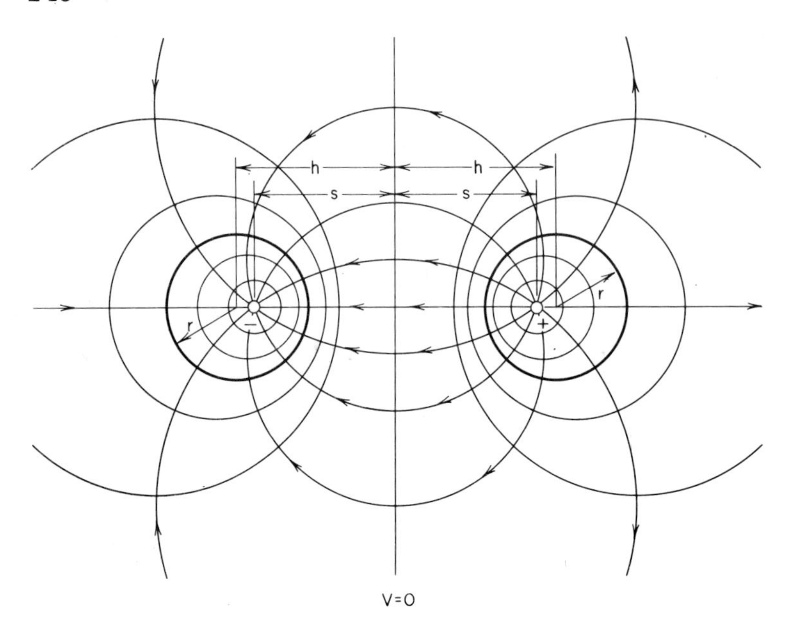

Fig.6.10. Field and equipotential lines around an infinitely long two-conductor line.

spacing $2h$ can represent the two wires. The field and potential distributions external to the wire surfaces are the same as if the field were produced by two infinitesimally thin lines of charge with a spacing of $2s$. The charge is not uniformally distributed on the wire surface but has a higher density on the sides that are adjacent.

The potential difference V_c between one of the conductors and a point midway between them is:

$$V_c = \frac{Q_L}{2\pi K} \log\left(\frac{h}{r} \pm \sqrt{\frac{h^2}{r^2} - 1}\right)$$

The capacitance per unit length C/L of the two-conductor line is the ratio of charge per unit length Q_L on one conductor to the potential difference V_{2c} between conductors:

$$\frac{C}{L} = \frac{Q_L}{V_{2c}} = \frac{\pi K}{\log\,[h/r \pm \sqrt{(h/r)^2 - 1}]} \text{ farads/meter}$$

$$= \frac{27.8 K_r}{\log\,[h/r \pm \sqrt{(h/r)^2 - 1}]} \text{ picofarads/meter}$$

where K_r is the relative permittivity of the material between the wires.

A single transmission line with a ground return brings about another form of capacitive coupling. Let the conductor radius be r and the height of the center of the conductor above the ground be h. Assume that the conductor has a positive charge Q_L per unit length and the ground is at zero potential.

The method of images is used to find the field and potential distributions. Thus, if the ground is removed and an identical conductor with a charge $-Q_L$ per unit length is placed as far below ground level as the first conductor is above, the problem is the same as for a two-conductor line. The second conductor, which replaces the ground, is the image of the upper conductor. The field and potential distributions for the single conductor line are illustrated in Fig.6.11.

The difference in potential between the single conductor and the ground is therefore twice the value of that for a two-conductor cable, or:

$$\frac{C}{L} = \frac{55.6K_r}{\log\left[h/r \pm \sqrt{(h/r)^2 - 1}\right]} \text{ picofarads/meter}$$

The surface charge density on the conducting ground plane is not uniform. It is at a maximum adjacent to the wire and zero at an infinite distance.

A schematic diagram of a simple unshielded two-conductor cable, such as

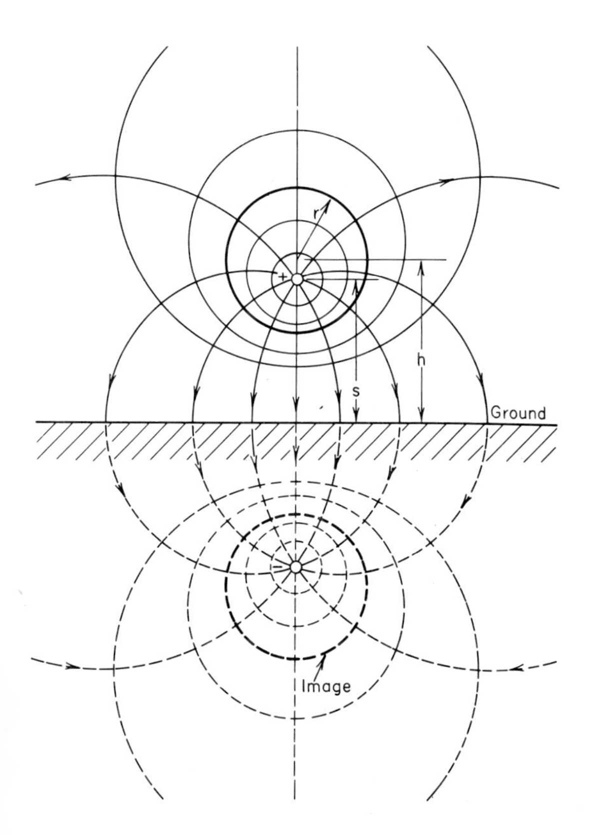

Fig.6.11. An infinite single conductor above ground with an electrical image.

might be used for two-array (pole—pole or half-Wenner) surveying, is given on Fig.6.12. This type of problem employing a grounded shield on the potential wire has been solved by Wait (1959c). The shielding does not prevent capacitive coupling, but the effects of the capacitive coupling are relatively constant and are therefore predictable. Wait found that at 3 Hz for a typical 1000-foot-deep diamond drill hole using wire with a capacitance of 5 nF the capacitive coupling would be on the order of $-3.2 \cdot 10^{-4}\ \Omega$. This is the correction factor to be applied as a percent decrease in mutual impedance in a given field problem.

The intercapacitive effects using a four-conductor cable can be appreciably large and complex; therefore, use of such cables should be avoided in surface, drill-hole, and underground IP work. The frequency method is sometimes said to be more prone to capacitive coupling than the pulse method, a statement which is true if the coupling phase angle is unknown, but essentially the same coupling difficulties exist using the pulse method.

Capacitive coupling can exist due to:

(1) Displacement currents from wire to wire or wire to ground, as discussed.

(2) Leakage currents from transmitter or receiver electrodes to receiver or transmitter wires. This leakage effect modifies the charge distributions affecting capacitance.

(3) A combination of these effects.

Figure 6.13 illustrates the condition described in (2) above. Madden and Cantwell (1967) point out that the capacitive coupling effect is negligible for the dipole—dipole and pole—dipole (three-array) configurations using expectable capacitance values. However, current leakage effects can be appreciable when using the Wenner, gradient, or Schlumberger arrays and can be as large as −2.0% at 3 Hz in extreme cases.

In order to minimize capacitive coupling, a receiver wire and a transmitter electrode should be well separated. Also the transmitter and receiver wires should be well separated, if it is necessary that they be in the same vicinity

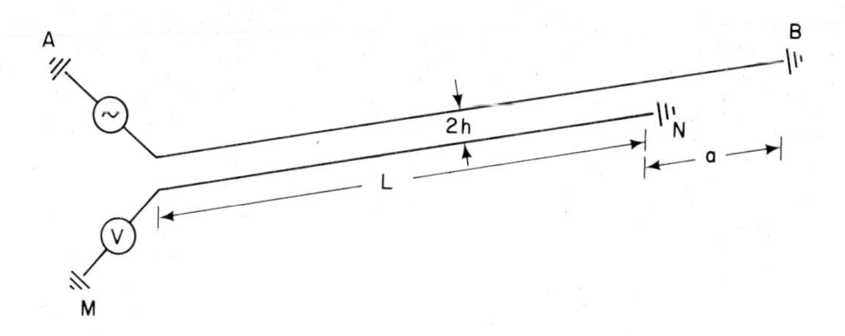

Fig.6.12. An equivalent circuit of interline capacitance, using the pole—pole array.

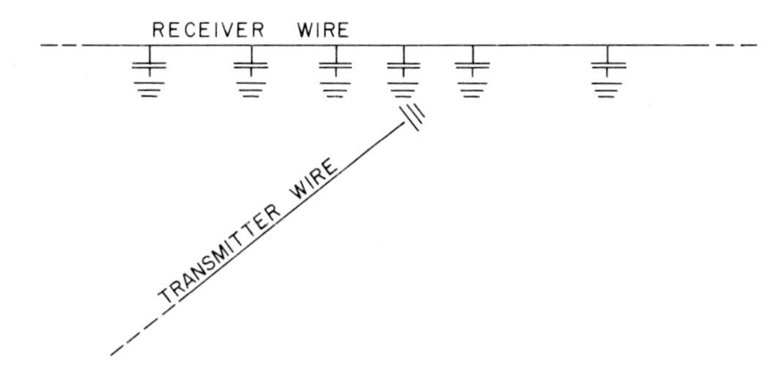

Fig.6.13. Capacitive effect with a wire due to leakage current from an electrode.

during a survey, particularly in a wet (conductive) environment. In a wet environment, taping of splices is especially necessary to minimize electrical leakage effects.

Cultural coupling effects

The 'fence effect' due to the presence of a nearby artificial conductor causes spurious IP anomalies primarily by leakage currents between transmitter and receiver circuits. The fence (or pipe line, or whatever artificial conductor is involved) acts as a good conductor between grounded points, but there are nonlinear or nonohmic electrical resistance effects at these grounded points where the current density adjacent to the grounded fence post is abnormally high. The effects of these leakage currents behave in much the same way as true polarization effects, and routine survey methods will not detect the difference.

Fence effects can be minimized by running traverses perpendicular to the fence, by insulating the fence, sometimes by using lower frequencies or currents, or, best of all, by cutting the offending fence or electrically or physically removing it.

ELECTROMAGNETIC COUPLING

In certain situations, the IP transmitter and receiver circuits behave like the primary and secondary windings of an ordinary electrical transformer. The primary circuit induces a current in the secondary circuit, the electrical induction effect being more pronounced at higher frequencies. Distance factors and low resistivities become important when interelectrode distances are within even a small fraction of the wavelength of electromagnetic radiation. This electromagnetic induction or 'EM coupling' causes spurious IP-like effects that are not due to natural polarization causes.

The EM-coupling problem

The problem of the electromagnetic inductive coupling of arbitrarily oriented lengths of wire on the earth has been treated at some length by Sunde (1949). He gives electromagnetic coupling as:

$$Z_{Ss} = \int_A^B \int_M^N \cos\theta \, P(r) \, dS \, ds + Q(r) \tag{81}$$

where Z_{Ss} is the mutual impedance between lengths S and s and θ is the angle between infinitesimal length dS and ds, as shown on Fig.6.14. In this relationship:

$$Q(r) = Q(A-B)(M-N) = Q(AM) - Q(AN) + Q(BN) - Q(BM) \tag{82}$$

and $P(r)$ and $Q(r)$ are functions dependent on the subsurface geometry and distribution of physical parameters. $P(r)$ is the electromagnetic coupling term, which can be written for a flat, electrically uniform earth as:

$$P(r) = \frac{i\mu\omega}{2\pi r}\left[\frac{1-(1+kr)\,e^{-kr}}{k^2r^2}\right] \tag{83}$$

in which k, neglecting displacement currents, is the propagation constant whereby:

$$k^2 = \frac{i\mu\omega}{\rho}$$

and:

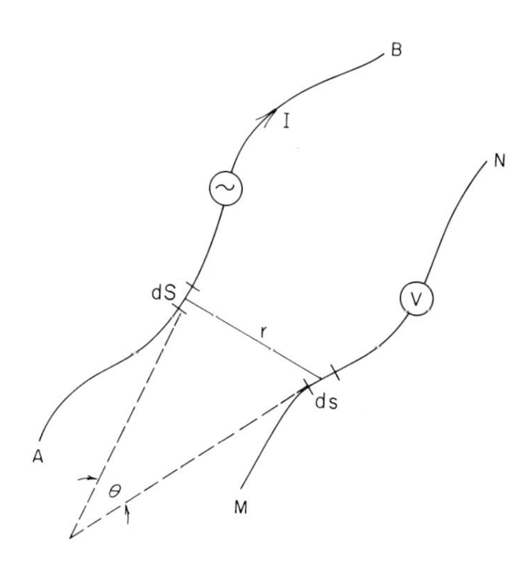

Fig.6.14. Elements used in Sunde's (1949) coupling equation. Wires extend from A to B and from M to N.

$I = I_0 e^{i\omega t}$

Mutual impedance Z_{Ss} of eq. 81 can also be written $Z(\omega)$, and the percent decrease in mutual impedance (PDMI) between two frequencies ω_1 and ω_2 can be written:

$$\text{PDMI} = \left[\frac{Z(\omega_2) - Z(\omega_1)}{Z(\omega_1)}\right] 100$$

where ω_1 is the higher and ω_2 the lower angular frequency. $Q(r)$ is the dc or resistive coupling term, and eq. 82 is essentially the same as eq. 19 for a homogeneous earth where $Q(r) = \rho/2\pi r$. For a layered earth, $Q(r)$ attains values that can be described as 'reflective coupling'.

EM coupling and phase

In order to explain phase relationships of EM coupling and the resulting effects on IP field work, it is convenient to use a frequency-domain phasor notation of electrical impedance. Here it is assumed that the ground is a linear system (Fig.1.3) with a transfer impedance $Z(\omega)$ such that:

$$Z(\omega) = \frac{V_p - V_t}{I_p} \tag{84}$$

where V_p is primary voltage, V_t is the total apparent polarization voltage measured at the output, and I_p is the inducing current input. Impedance consists of real (R) and imaginary (i) components

$$Z(\omega) = Z(\omega)\Big|_{R} + iZ(\omega)\Big|_{i} \tag{85}$$

Figure 6.15 shows phasor relationships using transfer impedance $Z(\omega)$ with electromagnetic coupling, with induced polarization, and with a combination of the two effects. These diagrams give an appreciation for the fact that while phase relationships between IP and EM coupling may look the same, the difference becomes important at larger EM coupling phase angles. At small phase angles, changes in the value of the mutual impedance coupling vector can be treated algebraically and the EM coupling component bears a proportional relationship with IP impedance. At larger phase angles, EM coupling and polarization relationships must be treated vectorially. The consequence of phase analysis in EM coupling is such that EM coupling cannot be completely compensated for unless the phase is known, and since coupling and polarization phase cannot easily be distinguishable, it may not be possible to make an adequate correction at large phase angles.

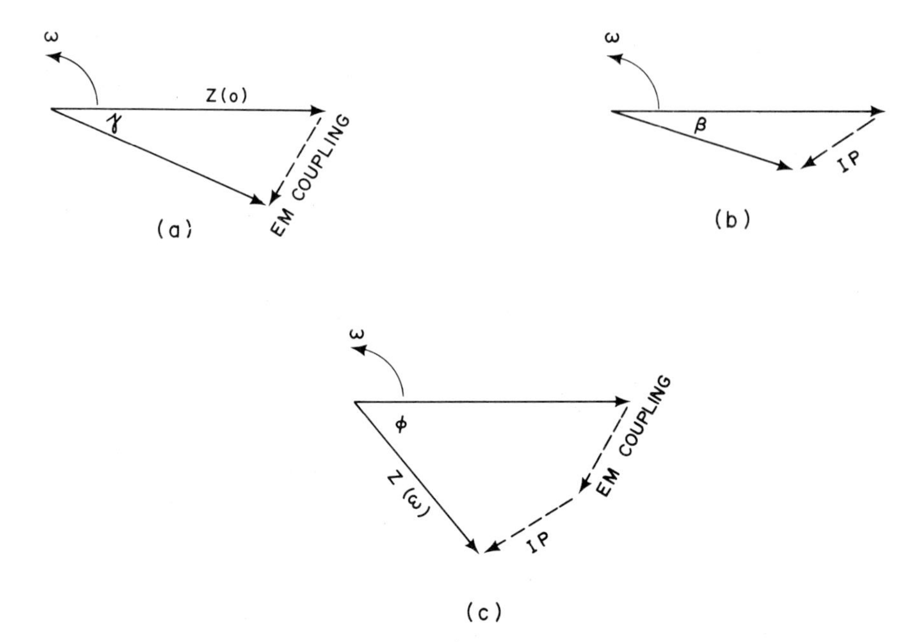

Fig.6.15. Phasor diagrams of transfer impedance showing vector components, (a) in electromagnetic coupling, (b) in induced polarization, and (c) a combination of the two effects showing the vector addition of EM coupling and IP values.

EM coupling relationships — correction curves

Solution of the coupling equation (eq. 81), involves integration of the first term, which is an infinite series:

$$P(r) = -\frac{\rho}{2\pi}\left[\frac{k^2}{2r} - \frac{k^3}{3} + \sum_{n=4}^{\infty} \frac{(-1)^n (n-1)\, k^n\, r^{(n-3)}}{n!}\right] \tag{86}$$

and this is a lengthy process. Evaluation of eq. 86 can yield both the coupling phase and the decrease in mutual impedance, and from this a correction of percent coupling effect at small phase angles can be obtained. Tables of calculated Z_{Ss} values are available (Millett, 1967) from which coupling correction curves can be derived. Coupling correction curves for a uniform half-space have been published by Kaku (1966) and by Geoscience, Inc. (1968). An example of phase angle and the percent coupling effect for the dipole—dipole array is given in Fig.6.16.

A series of nomographs will be presented which can be used as EM coupling corrections for IP field data, if phase-angle effects are not too large. It can be stated as a general rule that if percent EM coupling effects are less than 5%, a coupling correction can be safely applied to IP field data to give a correction with less than a one percent error.

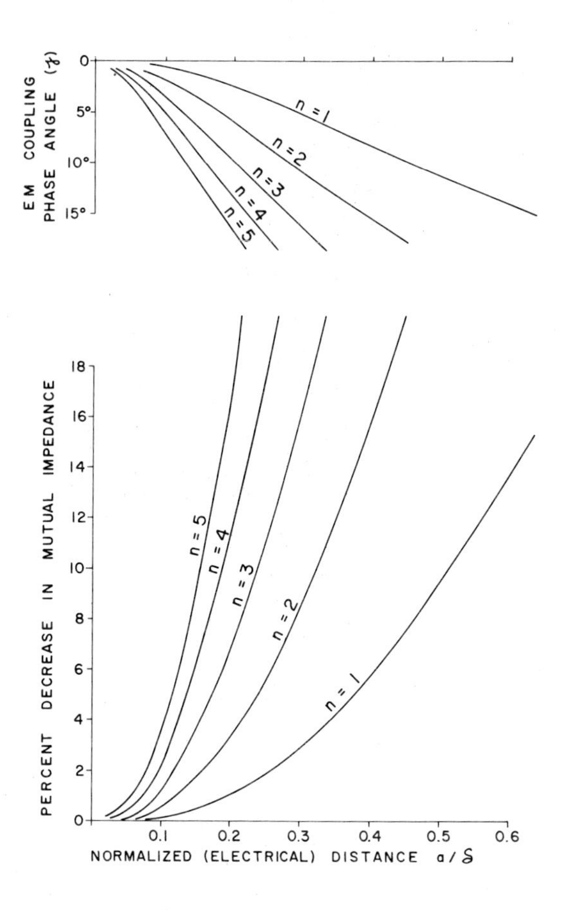

Fig.6.16. Out-of-phase and in-phase EM coupling effect curves for the collinear dipole—dipole array, showing phase lag and percent decrease in mutual impedance plotted against numerical distance a/δ.

The EM coupling problem has also been solved for two layers (Sunde, 1949; Hohmann, 1972), and N. Ness (written communication, 1961) has analyzed EM coupling solutions for two layers with different resistivity contrasts. The conclusion of these analyses is that if the EM coupling effect is not too large, less than 10%, the simple layered case can be treated as if it were a homogeneous half-space with less than 2% error.

EM coupling effects are positive, in the same sense as true polarization for an electrically homogeneous or layered earth when using the dipole—dipole array, but can be negative, for example, when using the gradient array. EM coupling effects may also be associated with anisotropic and inhomogeneous earth materials if low resistivities are involved. Cultural features, such as fences and pipe lines, may also contribute to an EM coupling effect.

EM coupling for small inductive effects using the dipole—dipole array

In the consideration of corrections of the EM coupling phenomenon, it is possible to normalize the theoretical results such that one specific geometric solution is applicable to a large number of physical situations. The nomographs shown in Figs.6.17 through 6.20 were originally developed by Ness in 1956.

For the dipole—dipole array, it is necessary to introduce the dimensionless parameter a/δ, which measures the ratio of the electrode spread length a to the skin depth δ. This 'numerical depth' is a measure of the importance of EM coupling in an IP survey. If a/δ is small, it means that the survey is being carried out on a resistivity and geometrical scale such that EM coupling effects are negligible (note that a/δ being small implies that δ is large). As a/δ increases, the EM coupling effects increase and at a very rapid rate. Because of this behavior and the manner in which δ depends on the physical parameters ρ and f, it is necessary to have actual values to predict the EM coupling effects for specific situations and subsurface structures.

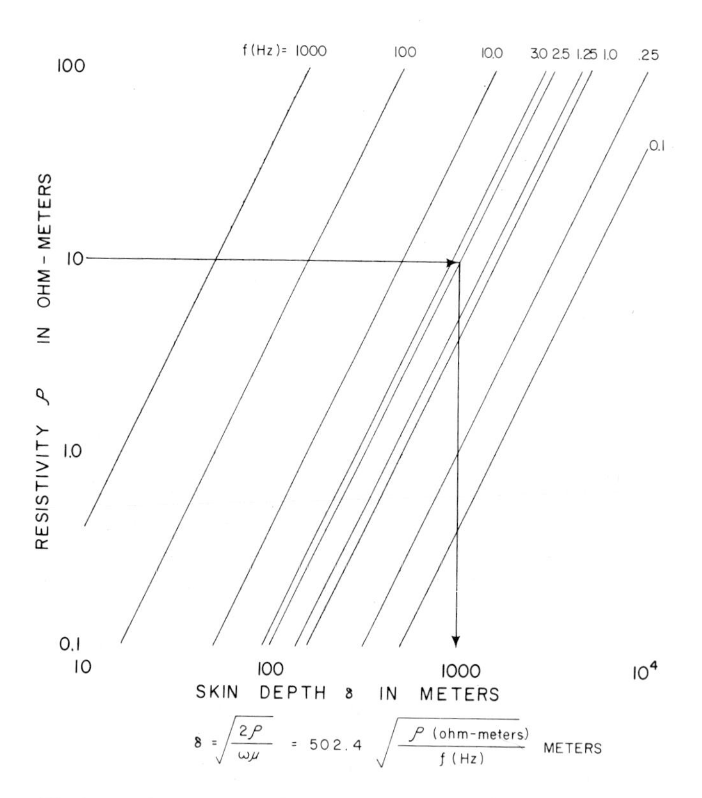

Fig.6.17. Frequency f plotted against resistivity ρ, in ohm-meters, and skin depth δ, in meters.

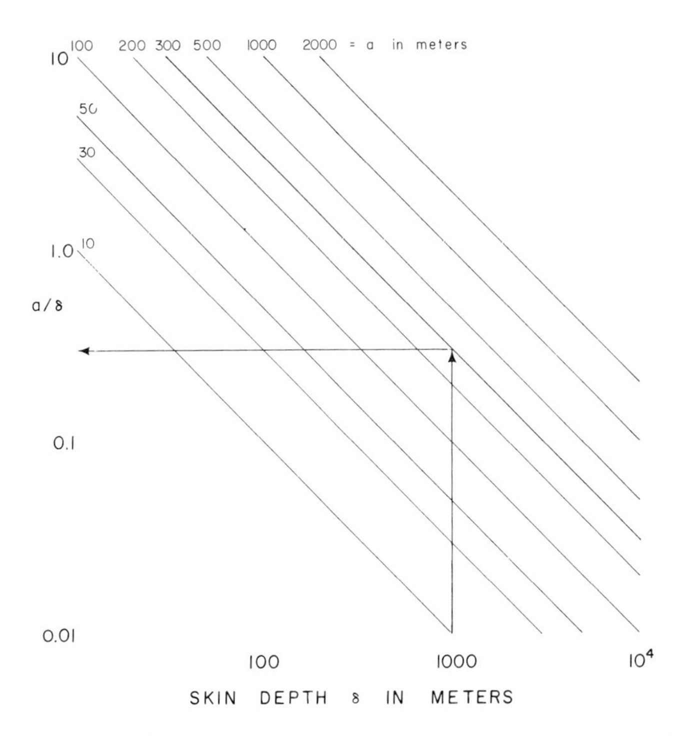

Fig.6.18. Electrode interval a, in meters, plotted against numerical depth a/δ and skin depth δ.

The simplest model of the earth is that of a uniformly homogeneous, isotropic half-space. The theoretical percent frequency effect for this model earth is presented in Fig.6.20 as a function of a/δ and n. It is seen that EM coupling effects increase extremely rapidly if $a/\delta > 0.2$ and are appreciable even if $a/\delta = 0.1$. It is also interesting that the effects are a sensitive function of the spread separation n. This section presents an example of the use of nomograms in determining the theoretical EM coupling effects. For $a/\delta = 0.20$, the phase angles associated with the received voltage relative to the current source are less than 20 degrees. Hence, if the percent EM coupling effect, or percent decrease in mutual impedance (PDMI), is insufficient to explain the PFE observed in the field at a particular location, it is possible to subtract the EM coupling effects linearly and obtain a residual that can be ascribed as being due to true polarization effects. For large a/δ, the phase angles are larger so that there is even less justification for the above procedure, since the phase of the signal measured should properly be taken into account, as discussed earlier in this chapter.

In general, IP prospecting should be carried out with $a/\delta < 0.1$, so that EM coupling effects are negligible relative to other sources of field data error. This EM coupling effect rule apparently holds for both a homogeneous and for a layered earth. A study by Ness of the EM coupling in fourteen layered situations indicates that:

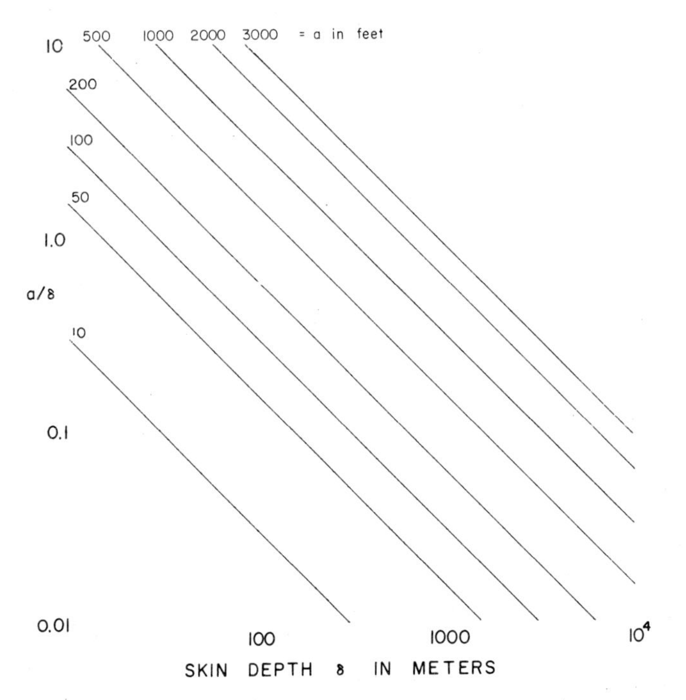

Fig.6.19. Electrode interval a, in feet, plotted against numerical depth a/δ and skin depth δ.

(1) When the total EM coupling effect is 5% or less, the true coupling value is nearly equal to the approximate calculated value, assuming a homogeneous earth. The differences are less than 1% frequency effect and generally less than 0.5%.

(2) For those cases with a higher resistivity overburden, the exact EM coupling effects are slightly larger than the approximate values calculated by assuming a homogeneous subsurface.

(3) For those cases with a lower resistivity overburden, the exact EM coupling effects are slightly smaller than the approximate values calculated by assuming a homogeneous subsurface. This fact tends to act in favor of a conservative interpretation.

Use of EM coupling curves for the collinear dipole—dipole electrode array

The data required are:

(1) Frequency f in cycles per second (Hertz).

(2) Dc apparent resistivity ρ in ohm-meters.

(3) Electrode spread length a in feet or meters.

(4) Electrode spread separation n in units of a.

Procedure:

(1) From Fig.6.17, determine δ, the skin depth in meters. Observe the

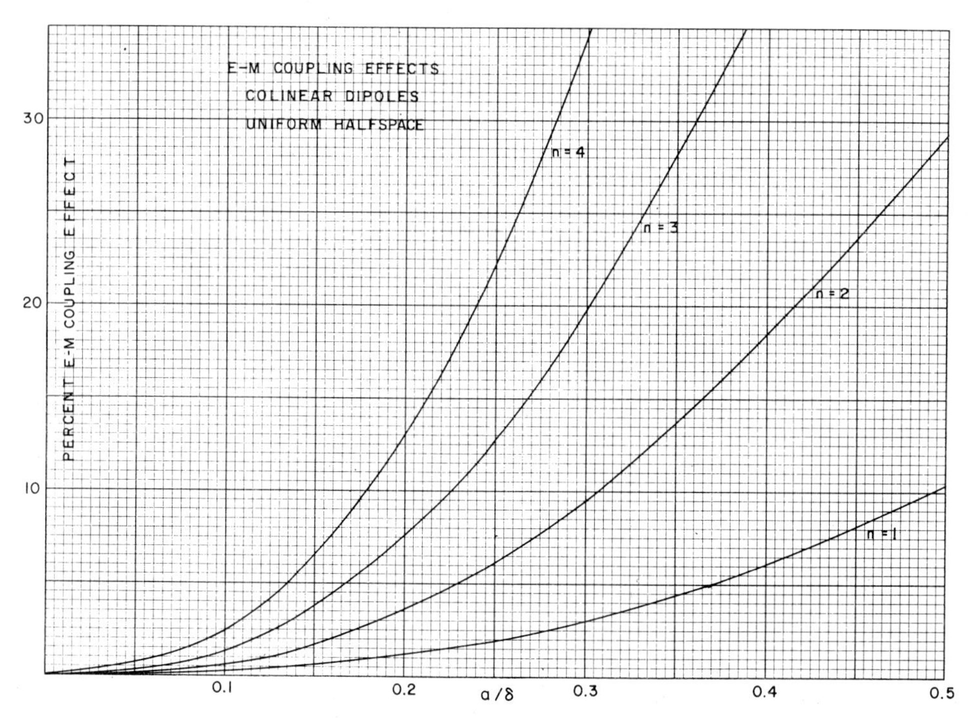

Fig.6.20. Percent EM coupling effects of collinear dipoles on a uniform half-space, shown as percent decrease in mutual impedance (PDMI) plotted against numerical depth a/δ.

intersection of a horizontal line drawn from the appropriate resistivity ρ value to the correct frequency f and then vertically to the δ value. This can also be computed directly from:

$$\delta = 502.4 \sqrt{\frac{\rho\,(\text{ohm-meters})}{f\,(\text{Hz})}} \text{ meters}$$

(2) With the a and δ values, use Fig.6.18 or 6.19 to determine a/δ, the numerical distance. Observe the intersection of a vertical line drawn from the δ value to the appropriate a curve and hence horizontally to the a/δ value. This can also be computed directly.

(3) With a/δ as determined above, use Fig.6.20 to find the percent effect due to EM coupling. Observe the intersection of a vertical line drawn from the a/δ value to the appropriate n curve and hence horizontally to the percent EM coupling effect.

Example:

$f = 2.5$ Hz $\qquad a = 300$ meters

$\rho = 10$ ohm-meters $\qquad n = 2$

f from Step 1, δ = 1,005 meters, and from Step 2, a/δ = 0.29, then from 'tep 3, percent EM coupling effect (PDMI) = 8.8.

Methods of avoiding EM coupling

As may be appreciated from the uncertainty which arises in attempting to compensate IP field data for large EM coupling effects (Fig.6.15c), the best way to treat the EM coupling problem is to avoid its effects rather than to apply corrections. This can be done by examining the skin depth δ of a homogeneous half-space, which was previously given by eq. 79 as:

$$\delta = \sqrt{\frac{2\rho}{\omega\mu_0}}$$

where ρ is resistivity, ω is angular frequency, and μ_0 is magnetic permeability. The electrical depth of a half-space, normalized in terms of electromagnetic wave penetration, can be defined as the ratio, a/δ, where a is the electrode interval. In order to decrease the EM coupling effects, it is necessary to decrease the electrode interval a, the electrode separation factor n, or the frequency.

Note from eq. 81 that there is a $\cos\theta$ term between wire segments entering into the coupling (Fig.6.14). By choosing an array where $\cos\theta \rightarrow 0$, such as the pole—dipole configuration on Fig.6.21a, EM coupling can be minimized. One can also use the perpendicular dipole—dipole array shown in Fig.6.21b rather than the collinear dipole—dipole arrangement. However, even when wire pairs are at right angles, contrasts between materials of different resistivity or polarization distort the electromagnetic radiation pattern and can cause transient polarization-like response. This type of response has been called electromagnetic reflection.

Detection and identification of EM coupling

A rough means of detecting the presence of EM coupling in field data is to observe and measure the shape of the decay curve during the first second of recording or to view the frequency or phase spectrum at frequencies greater than 1.0 Hz. From the resistivity data of a survey area it is possible to predict the EM coupling, then remove this effect from field results to obtain the true IP response. Also, phase spectra appear to be useful in predicting EM coupling, and K.L. Zonge seems to have devised a method for separating IP and EM coupling effects.

Figure 6.22 shows characteristic EM coupling response curves. The EM coupling effect can be crudely observed in conventional field data by making measurements at geometrically spaced time or frequency intervals. If the data show that response values increase at earlier times or higher frequencies it is likely that EM coupling is present. For example, PFE readings can be made at 0.1, 0.3, and 1.0 Hz, and if the response is larger between the two higher frequencies than between the two lower values, EM coupling can be suspected.

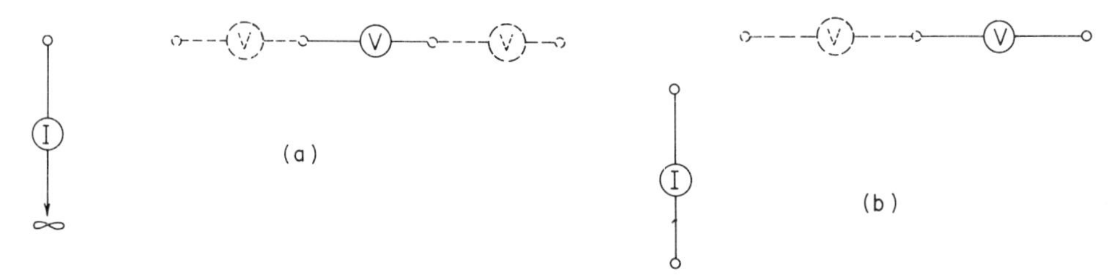

Fig.6.21. Plan view of arrays that tend to minimize electromagnetic coupling: (a) the pole—dipole or three-array, and (b) the perpendicular dipole—dipole array.

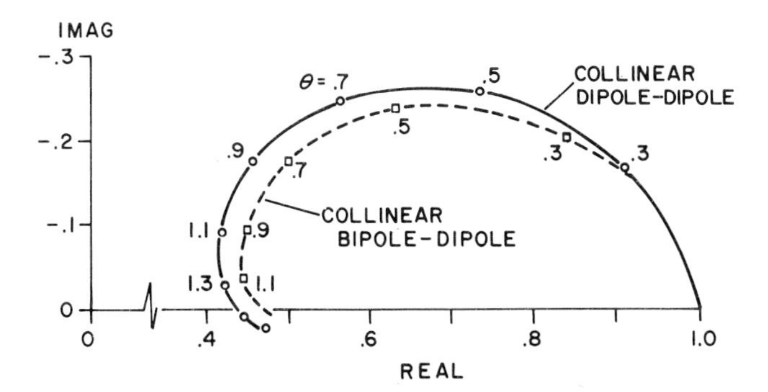

Fig.6.22. Homogeneous earth collinear dipole—dipole (solid line) and collinear bipole—dipole (dashed line) EM coupling; $\theta = (\omega\mu/\rho)^{1/2}$.

If there is a strong departure of a recorded frequency spectrum or decay curve from a logarithmic response at higher frequencies or earlier times, EM coupling is likely present. However, for routine surveys it may be just as easy:

(1) To precompute the EM coupling frequency or time cut-off value, finding where EM coupling effects will exceed 5% frequency effect or 20 milliseconds chargeability.

(2) To adjust the frequencies or recording times accordingly, so that the larger coupling effects are avoided.

(3) To correct the field data for EM coupling where the effect is less than 5% FE. Where coupling effects inadvertently exceed 5%, the data should be questioned or discarded.

EM coupling can be avoided more easily in the time domain than in the frequency domain, merely by providing a time delay between the instant the inducing current is turned off and the time the decay voltage is observed. This delay time is standardized at 0.45 s on the Newmont receiver, but it may necessarily use a delay time as large as 1.0 s in unfavorable circumstances. Figure 6.23 is a comparison of time and frequency filtering of the delay time step-function of a time-domain IP receiver.

The ratio *L/M* (Fig.5.13) taken from a Newmont receiver can also be helpful in detecting EM coupling. Ratios greater than 0.8 indicate the shorter time-constant decay to be expected when EM coupling is present.

Under special circumstances in field surveying it may be worthwhile to use a different array, either one of the right-angled arrays suggested in the previous section or in the case of the collinear dipole—dipole array, to try a parallel dipole—dipole array, as illustrated in Fig.6.24a. This array will give a negative response if EM coupling is present. Also, the *a* spacing can be increased and *n* decreased, as shown in Figs.6.24b and 6.24c, giving the same depth of exploration but less coupling.

Unfortunately, the phenomenon of electromagnetic coupling often limits the useful application of the conventional IP method to very low frequencies and longer times to the point where noise problems start entering into the situation. Wherever possible, EM coupling effects should be avoided.

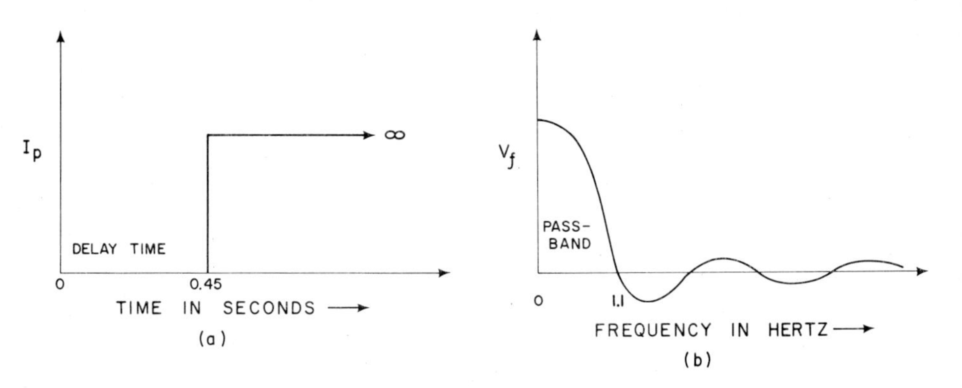

Fig.6.23. Comparison of the time-domain step-function inducing current waveform (a) with a comparable frequency band-pass filter waveform (b).

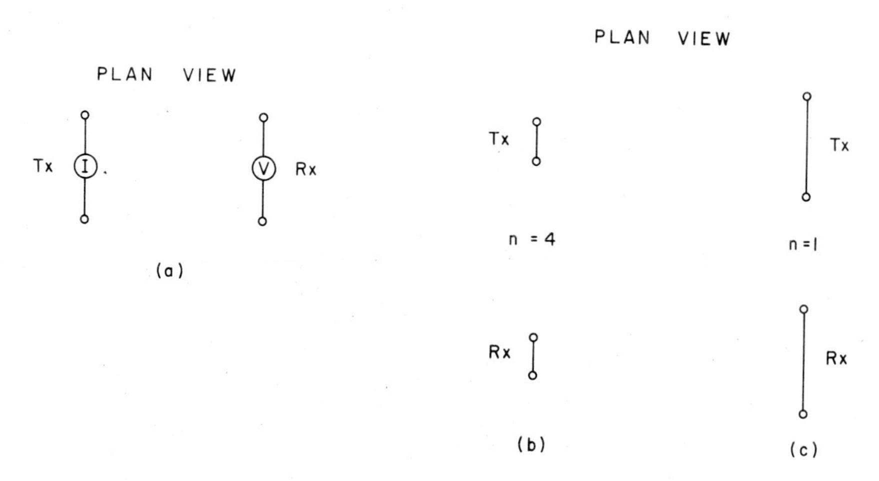

Fig.6.24. Plan views of electrode arrays which can decrease EM coupling effects: (a) the parallel dipole—dipole array which gives acceptable survey results but with a negative EM coupling factor and (b) and (c) two collinear dipole—dipole layouts; (b) has a smaller *a* spacing and larger *n* than (c) and (c) has a lower EM coupling effect than (b) and also a much better signal-to-noise ratio.

Chapter 7

IP FIELD SURVEYING

Induced polarization field surveying is usually a rather complex measurement operation involving several men, perhaps two or three vehicles, and a large amount of expensive, specialized equipment. It is necessary that the IP field crew be organized so that work will progress efficiently, producing a maximum amount of good data without unnecessary labor on the part of the crew.

The organizational chart of Fig.7.1 is typical of IP field exploration parties operating in North America. The crew chief is usually a senior geophysical technician or a geophysicist, acting under the supervision of a regional or chief geophysicist. The crew chief is responsible for all field operations, including specifically checking out the survey layout, taking receiver readings, care of equipment and personnel, daily calculation and plotting of field results, relations with the local geologist or client, and making a field interpretation of data. Other duties are frequently included, making the crew chief a busy man and certainly the personification of the IP crew.

The IP transmitter operator is a field person who requires some on-the-job

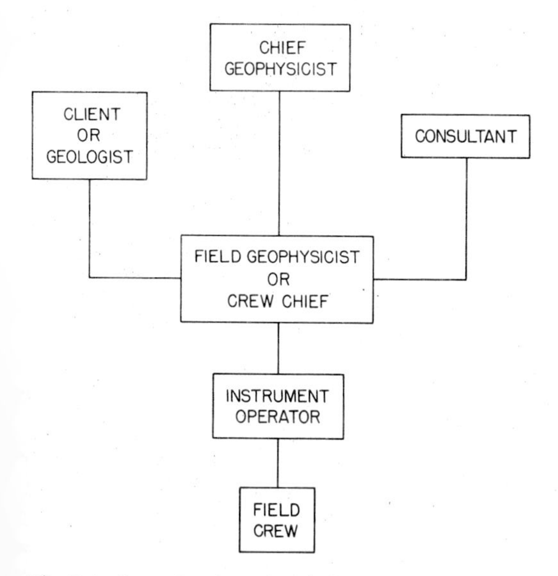

Fig.7.1. Organization chart of a typical IP field party.

training, perhaps just a few days. Workers taken on locally as wire-pullers or 'electrode emplacement engineers' need only be willing, alert, and able to follow multistage instructions.

FIELD PROCEDURES

Each IP field survey can be considered to be a somewhat different applied research project carried out under varying conditions, and field work must be planned accordingly. Initial plans for an IP field survey are usually made in some distant exploration office and are based on the best information available about the property. Plans must take into account the nature of the survey area and the exploration target, the desired coverage, capabilities and limitations of field equipment, and the field crew. After preliminary discussions with knowledgeable geological personnel or the client, the regional geophysicist or his representative usually visits the property with the field crew in order to see the setting first hand and to be present when the first data are taken. Often initial plans must be modified in the light of unforeseen field conditions, and it is desirable to have expert assistance before revising plans.

It is good practice to have the field chief report periodically to the geophysical supervisor. The field chief must always be available for communications from above. Frequent guidance is desirable during initial stages of a project. As a program becomes routine, less contact is necessary. Of course, the exploration manager is continually anxious to receive all of the latest survey information so that he can extend or modify coverage or take the utmost advantage of a potential mineral discovery. Thus, data must be continually computed, plotted, and field interpreted.

Logistics of field equipment is an important matter, and availability of local transportation and parts facilities must be borne in mind in planning for providing extra equipment and spare parts. If a crew could be held up for two or three days for lack of a component, then it is best that a spare be immediately available with the crew. Since much of the supplementary IP field equipment is expendable with use, it is necessary to lay initial plans contingent on the anticipated duration of surveying in the area.

Equipment list

Before embarking on an IP field expedition, good planning procedure dictates that a list of standard equipment be made, with a supplementary list of accessories, spare parts, tools, and expendable items. Table VII is a typical (but hypothetical) list of field items for an IP survey of short duration. For proper preparation for a field surveying expedition, all eventualities should be considered and preparations made beforehand.

TABLE VII

IP survey equipment list

Basic equipment	Spare equipment
Voltmeter-receiver	Spare components; extra batteries
Current-sender; dummy load	Spares
Motor-generator; spark plug, gasoline, oil	Extra spark plugs
Wire and reels	Extra wire and reels
Porous pots; copper sulfate	Extra pots
Current electrode materials, aluminum foil, water cans, salt	
Voltmeter-ohmmeter	Extra batteries
Walkie-talkie radios	Extra batteries
Two off-road vehicles	Extra gasoline
Tape, pliers, wire-strippers	
Equipment manuals	
Project files	
Note paper, data sheets, plotting paper	
Compass, maps, surveyor's tape	
Fire extinguishers for vehicles	
Powder-type extinguisher for motor-generator	
First-aid kit, including snake bite kit	
Personal items and pack sacks	

Transportation

Since field equipment and the entire survey are often designed around the type of available transportation, initial thought must be given to the portability of equipment and the mode of transport. If the vehicle's motor is also a power source, a reliable power-train and throttle governing system become necessary.

Survey layout

Previous planning for the IP field survey would have considered the direction of the geologic 'grain' and the probable depth and size of a polarizable target. IP survey traverses are usually laid out at right angles to the geologic strike, using an array with the desired resolution and an electrode interval that is no less than the desired optimum depth of exploration, with the distance between reconnaissance survey lines equal to as much as twice the electrode interval.

It is often convenient to separate the IP survey into an initial reconnaissance phase during which anomaly discovery is sought and a detailing phase during which data are gathered from which the size, shape, depth, and other details of a discovered polarizable body can be interpreted.

Too great an electrode interval may miss a narrow polarizable body and too small an interval may not 'look' deep enough. Vozoff (1958) has pointed out that maximum survey information is obtainable only by having transmitter and receiver locations virtually everywhere on the ground surface. Since this excessive amount of surveying is impractical, one must plan an optimum layout for the most probable surface conditions, taking care that no reasonably discoverable polarizable body could be overlooked. For the sake of survey efficiency and economy, the survey interval is a type of filter which can be adjusted for size and physical property of the target. In order to gain a maximum amount of information about the subsurface, an experienced geophysicist in consultation with a knowledgeable geologist should lay out the IP survey in a new area.

Sometimes it is necessary to lay out survey lines in directions other than normal to the geologic strike. For example, if extensions of a thin, tabular, polarizable body are being sought, it may even be necessary to run a traverse in the strike direction over the body so that the IP anomaly will be better defined. In very rough terrain, the IP traverses can be laid out along topographic contours or along ridge or drainage trends, saving surveying time in the field. The effect of fences is best minimized by crossing them at right angles or nullifying their effects by cutting or removing with permission. Property boundaries sometimes must be paralleled and not crossed, although reciprocal agreements can usually be made with the owners.

Permission to survey

Because IP surveying is costly, some exploration companies believe that property should be acquired or otherwise controlled before surveying is done. However, land acquisition is also costly, and so other companies have carried out reconnaissance work on speculation on government land and even on property held by others. The topic of mineral rights and the right to trespass on unpatented land or land with certain ownership restrictions is a legally complex topic beyond the scope of this book.

As the world becomes more populous, the demand for additional raw materials causes an increase in exploration activity. But consideration for the condition of the land also increases, and there is a concern on the part of many people that activities, such as use of off-the-road vehicles, line cutting for laying out surveys, and digging holes for electrodes must be reduced. Therefore, it behooves all people involved in mineral exploration to improve their public image and to become exemplary campers and outdoorsmen. Gates must be closed, wire picked up, holes filled, and other good housekeeping tasks taken care of before leaving the exploration scene. On the next visit, our presence will be more welcome.

THE CURRENT-SENDING SYSTEM IN THE FIELD

After an area has been selected for IP surveying and a general field plan has been agreed upon, the site is assessed by the field crew for line layout and selection of the first generator station location. Since the motor-generator is not easily portable, it is usually left at one location and the line is surveyed, that is, IP data are taken with the more mobile voltmeter-receiver.

The generator station

A generator station is selected because of its accessibility, for example, so that vehicles can drive to it if they are being used for the survey work. Also, the generator should be centrally located in the layout of current electrodes, so that wires can be strung out in both directions. If the generator station is located near the center of the spread, less wire is used, there is less chance of leakage coupling, and there is lower power loss in the wire. As an example, it frequently is a routine procedure when using the dipole—dipole electrode array employing the roving receiver method to run a set of six dipoles with the motor-generator and transmitter set up at the center of the six-dipole (seven-electrode) spread.

All IP field equipment would be initially transported into the generator station and from this point persons would dispatch themselves to prepare the current electrodes, perhaps pulling a current wire as they go. Two separate wires should not be pulled from the same reel or they may tangle. After the electrodes are prepared at the seven locations and the current wires attached, the crew chief or transmitter operator checks the resistance between each current dipole electrode pair, either with a pocket ohmmeter or by using the transmitter itself. If the current electrode resistances are within acceptable limits, the survey can proceed, otherwise one or two persons must go to the high resistance electrode location and improve (lower) the resistance by adding water or yet another individual, electrically parallel electrode.

At the generator station, the motor-generator is placed about 15 meters from the transmitter current control unit in a level, sheltered spot. Care is taken that dirt, gravel, or grass cannot be sucked into the air intake or flywheel housing and that the exhaust will not ignite nearby dry grass. The gasoline can is set at a safe, convenient place nearby.

Electrode preparation

One object in emplacing a current electrode is to give the nearby ground a minimum resistance so that a maximum amount of current can be passed through it. In order to achieve this condition, the current electrode should have a fairly large surface area in good electrical contact with the nearby moist disturbed ground. To serve this purpose, multiple driven stakes are used, also buried metal plates or sheets of aluminum foil in shallow trenches. To make the current electrode's electrolyte more conductive, common salt is usually added. Digging, blasting, or mud capping may be used to provide the proper low-resistance interface environment between the metallic electrode materials and the subsurface moisture.

Electrode resistances become lower with time if they are prepared with an abundance of conductive salt water. This general rule holds for all types of current electrodes, be they metal stakes, plates, or foil. Thus, electrode positions are often prepared the previous day and allowed to soak overnight.

The current-carrying wire from the transmitter is usually fastened to the electrode by twisting with the electrode material so that a strong pull will detach the wire and allow it to be reeled in. The wire is attached without a clip that might snag on bushes.

If the current wire is long, more than a hundred meters, there is an advantage to having a mechanical or motor-driven reel device at the generator station to expedite the picking up of wire and departure to the next site after a line is surveyed. More than about 500 meters of wire cannot be readily pulled through underbrush without first clearing a survey line. Survey wire should have a tensile strength of more than 300 pounds so that it can be pulled without breaking. Current wire usually has a resistance of 5–20 Ω/100 m and typically is a tinned, multi-strand copper-coated steel wire in a vinyl sheath. For proper splicing of this high-tensile strength wire, it is first necessary to fasten the ends together physically by tying a firm knot, then to make the electrical splice.

Review of theory of current electrode layout

In electrochemistry, electrode resistance refers to the electron transfer resistance effect near an electrode, such that the electrode becomes 'polarized', requiring an additional voltage to force current through the circuit. In IP and resistivity field work, there is only an approximation analogy between characteristics of the field electrode and the true polarization resistance phenomenon of the electrochemical electrode. 'Electrode resistance' in IP and resistivity field work means the amount of resistance contributed by the undisturbed ground within a relatively short distance from the electrode. Electrode resistance is the resistance contributed to the current circuit by the undisturbed earth resistance within, say, 5% of the total interelectrode interval.

Although for a homogeneous, isotropic subsurface Ohm's law, eq. 13, is obeyed, a plot of the resistance between two electrodes is not linear with distance. From eq. 18 a graph can be drawn of the total percent resistance between current electrodes A and B using the relationship:

$$\frac{V}{I} \doteq \frac{\rho}{2\pi}\left(\frac{1}{r_A} + \frac{1}{r_B}\right) \tag{87}$$

which is shown as Fig.7.2. Near electrodes A and B, the resistance gradient is very steep, and this is the electrode resistance region with which we are concerned. The near-electrode resistance can be lowered, but only by expending effort. The resistance-vs.-distance diagram of Fig.7.2 was derived by assuming single hemispherical current electrodes at A and B whose radii are 1.0% of the interelectrode distance AB. Equation 87 shows that electrode resistance is directly proportional to the resistivity of the ground.

Electrode resistance

The concept of a current electrode having an appreciable resistance is artificial, inasmuch as the electrode itself and the surrounding disturbed ground in fact have a very low resistance. However, the term 'electrode resistance' is useful if it is understood to refer only to the amount of resistance contributed by the zone near the electrode.

For the region near electrode A, the distance r_A is small and r_B is large. Thus, the second term of eq. 87 can be ignored and the resistance of electrode A can be arbitrarily defined as:

$$R_E = \frac{V}{I} = \frac{\rho}{2\pi}\left(\frac{1}{r_A}\right) \tag{88}$$

In the field and according to eq. 88, an individual current point electrode tends to act hemispherically. Thus, the electrode resistance must vary inversely

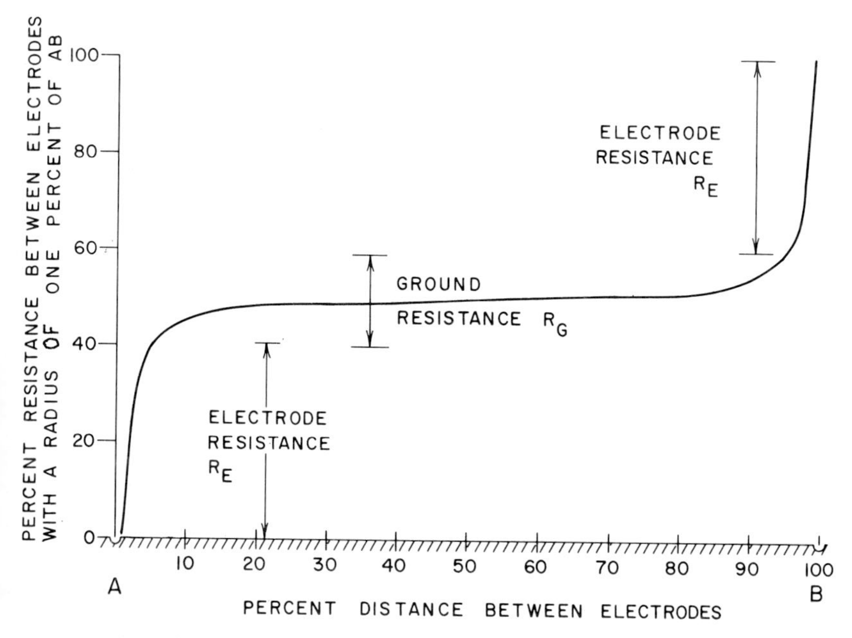

Fig.7.2. Plot of potential between two current electrodes.

with the cube of the radius of the water-soaked or disturbed volume which implies that in order to lower R_E by a factor of two, it would be necessary to soak or disturb a volume eight times larger than the original condition. In order to avoid this excessive labor consider next a flat, thin, circular electrode instead of a hemispherical one and it is then logical to look at the electrode resistance problem as an areal matter rather than a volumetric one. The area in contact with moist ground using the horizontal, flat electrode would be half that of a hemispherical one of the same radius but only the upper, dry soil layer need be disturbed.

Instead of a single current electrode now imagine two, three, or four smaller electrodes spaced in a regular pattern so that there is no overlap of the disturbed, wetted ground surface. For a current electrode pattern of n such electrodes, all electrically connected in parallel, superposition requires that their effects be additive, such that the group of n individual electrodes has a resistance:

$$R_E = \frac{\rho}{n2\pi}\left(\frac{1}{r_a} - \frac{1}{d}\right) \tag{89}$$

where d is here taken as 5% of the distance AB between a current electrode pair and r_a is the radius of individual current electrodes.

Equation 89 implies that by doubling the number of individual electrodes their combined resistance is halved. Each individual electrode takes labor to prepare so that it may make sense to put in two or three electrodes, but beyond that a point of diminishing returns is soon reached. The 'improvement factor' acting to lower electrode resistance varies inversely with the number n of electrodes.

When using multiple electrodes as described in the previous paragraphs, the radius of the cluster of individual electrodes should not be greater than about 5% of the total distance AB to the center of the adjacent electrode cluster. Resistivity calculations are based on the interelectrode distance, and the calculated resistivity would be in increasing error by the ratio of cluster-diameter to the distance AB. For this reason, it is sometimes practical to emplant current electrodes in a line perpendicular to the traverse, rather than a cluster, if the distance between members of a cluster of electrodes is an appreciable fraction of the distance AB.

It is possible to measure an effective stake resistance by using three stakes in three simple two-stake circuits and solving the resulting three resistance equations for the three unknown resistances. This method ignores the relatively small effect of ground resistance and distances between stakes.

The sender in operation

During the IP survey, the transmitter-sender is usually operated only periodically on command from the voltmeter-receiver operator. Communication may be visual or by sound, voice, radio, or telephone using the transmitter cable. Timed current transmissions can be made on a prearranged schedule, particularly when using the time-domain method or frequency-domain method or the sophisticated phase system where one consecutive group of waveforms is employed. With some surveys, such as the gradient array, the transmitter can be left on continually and unattended during field measurements. In using the transmitter in this fashion, operating costs are increased but they are offset by the requirement for fewer personnel.

Of course, the wire conductor between current electrodes must be opened for receiving (or have an infinite resistance) or an IP effect will be shorted out.

The transmitter operator's job is not difficult and is becoming less so with improvements in field equipment. Since this job is often boring, particularly for an alert person, simple but important mistakes can be made unless a good record is kept of the transmitted sequence. Typical data sheets for an IP transmitter and receiver operation are given as Fig.7.3.

McPHAR GEOPHYSICS, INC.

STANDARD TX-IP SURVEY DATA

Operator ________ Date ________

Property ________ Line ________

TX LOC ______ CAL ______ TX LOC ______

RX LOC	CURRENT		RX LOC	CURRENT

TX LOC ______ CAL ______ TX LOC ______

RX LOC	CURRENT		RX LOC	CURRENT

TX LOC ______ CAL ______ TX LOC ______

RX LOC	CURRENT		RX LOC	CURRENT

(a)

McPHAR GEOPHYSICS, INC.

Standard I. P. Survey Data

OPERATOR ________ DATE ________

PROPERTY ________ LINE ________

TX. LOC. TIME CAL.

RX LOC	ATT	A C VOL	I	F E	CORR / F E	PA / 2π	M F

TX. LOC. TIME CAL.

RX LOC	ATT	A C VOL	I	F E	CORR / F E	PA / 2π	M F

TX. LOC. TIME CAL.

RX LOC	ATT	A C VOL	I	F E	CORR / F E	PA / 2π	M F

(b)

Fig.7.3. Typical data sheets for IP transmitter (a) and receiver (b) operation.

After the current electrodes are put in and their resistances are found to be at an acceptable level, preparations are made to survey the traverse with the voltmeter-receiver. The duties of crew members change for this operation, and different tasks are performed than those while the transmitting system was being established. But before voltage measurements are made it is usually necessary to test the performance of the receiver and the transmitter by calibration.

Field calibration procedures

There is a need to compare the voltage measurement of an IP voltmeter-receiver against a known standard value so that IP readings from different areas or surveys using different equipment can be compared on an equivalent basis. The transmitter must also be periodically calibrated against a known current standard.

The receiver calibration procedure can be accomplished either by using an active, specially designed voltage source fed into the receiver or by measuring the voltage drop across a passive circuit element inserted in series in the transmitter and current electrode circuit. The active signal source calibration method can be used in the laboratory or in the field to check the performance of the voltmeter-receiver. The calibration instrument is a signal generator with a known, consistent waveform having a precisely controllable voltage amplitude and frequency.

A second IP equipment calibration method is more generally used in the field, and although it has the advantage of checking the performance of both the voltmeter and transmitter system, it cannot necessarily isolate the malfunction to either. In the frequency method, the field calibration circuit shown in Fig.7.4a is usually a purely resistive circuit element across which the voltage should be the same at different frequencies, if the same amount of current is passed through the circuit. The calibration circuit element is usually contained within the transmitter or receiver instrument case and is accessible by terminals on the control panel. If the voltage difference found in this calibration procedure is small, the resulting value may be applied as a correction factor (calibration factor) to the IP data for that area or to the data from that current dipole. In the frequency method, the need for this correction may be due to:

(1) Amplifier drift or small changes in filter characteristics in the voltmeter.

(2) Inadequate current control in the transmitter. This faulty control could also be caused or accentuated by (a) excessive electrode polarization

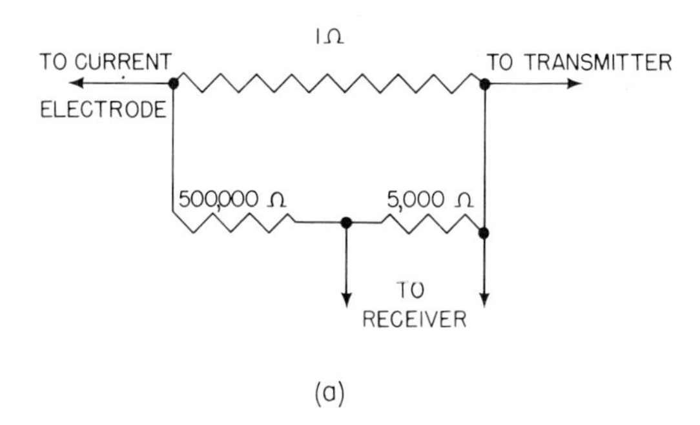

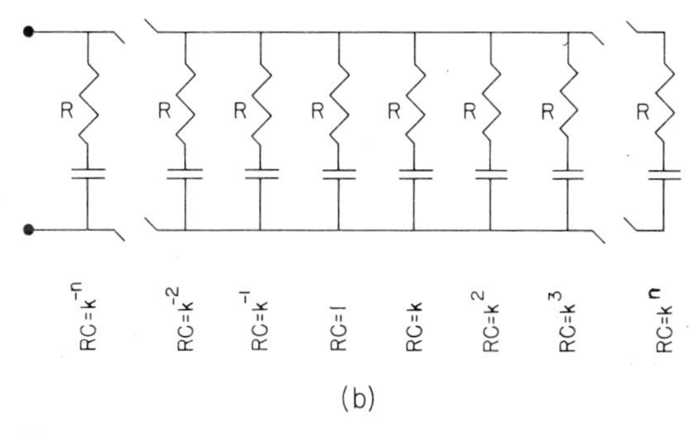

Fig.7.4. IP calibration circuits. (a) Circuit for calibrating frequency equipment. The voltage divider circuit is designed to have an impedance and voltage level similar to that of the ground. (b) A time-domain resistance-capacitance ladder network to simulate and therefore calibrate polarization response.

effects at the current electrodes or (b) polarizable material in the near vicinity of the current electrodes.

In the frequency method, an ideal IP transmitter will have perfect current control and will not contribute to a calibration voltage difference. Similarly, an ideal drift-free IP voltmeter will not have a calibration difference.

A time-domain IP calibration circuit is shown in Fig.7.4b. This is a ladder network or distributed circuit element that can be designed to approximate a logarithmic voltage response with time, simulating the polarization response of the ground. By using different selective values of components in the ladder network, chargeability values of zero, 10, 50, and 100 can be synthesized. Of course, the zero chargeability calibration circuit would be much the same as in Fig.7.4a. If desired, the *L/M* value can also be calibrated using a circuit similar to that of Fig.7.4b. In field equipment in which there is an electrical linkage between transmitter and receiver, such as described in Chapter 5, calibration is continuously accomplished by the signal comparison.

During the field calibration procedure, it is a good idea to separate the voltmeter physically from the generator and transmitter units. Electromagnetic noise transmitted from the generator and transmitter attenuates rapidly with distance. Also unwanted ground current will be less effective with distance. Caution is necessary during the calibration procedure because of danger to personnel and equipment from the nearby high-voltage, high-current transmitting system.

Even though some IP field instruments are remarkably accurate, a periodic calibration check is necessary to satisfy the cautious geophysicist. Calibration tests are proof of the quality of IP field data, and results of calibration tests should be recorded in the field notes.

The porous-pot potential electrode

After calibration, the field crew starts taking voltage readings. The actions of the transmitter and receiver systems are coordinated. The receiver circuit consists of the voltmeter-receiver, the potential wire, porous-pot electrodes in contact with the ground, and the electrode and ground resistances. This simple circuit is similar to the one for the current circuit illustrated in Fig.5.2.

The porous pots are emplaced by digging a small hole with a hand trowel or similar tool, perhaps filling it half full of water, and firmly emplanting the porous-pot electrode so that it is in good physical contact with the moist earth at the bottom of the pot hole. A section through a porous-pot electrode is illustrated in Fig.7.5.

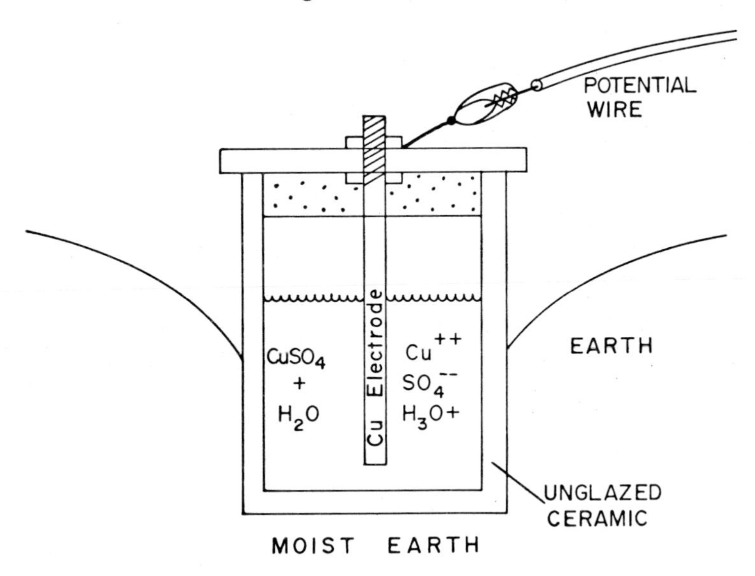

Fig.7.5. A cutaway section through a porous-pot potential electrode.

Porous pots should not be emplanted too close to the current electrode position because of possible leakage currents with resulting undesirable capacitive coupling effects. Also, the high current densities near the current electrode disturb the electrical equilibrium of the ground and cause a drifting potential at the porous pot which can be bothersome at very low frequencies or when measuring direct current.

The sum of pot resistance R_P and ground resistance R_G between potential electrodes at M and N is:

$$2R_P + R_G = \frac{\rho}{2\pi}\left(\frac{1}{r_M} + \frac{1}{r_N}\right) \tag{90}$$

As with current electrodes, it may be necessary to lower the pot resistance R_P by digging down to moist ground and adding water. Much higher resistances can be tolerated by the high impedance voltmeter-receiver than by the current sender, so that generally very little effort need be expended in preparing a potential electrode site. Nevertheless, it is a good idea to measure the resistance of the potential circuit to assure that it is always within acceptable limits. If the resistance is too high, more water should be poured into the pot holes to lower the pot resistances, the holes should be dug deeper, or the high-resistance hole should be abandoned and the pot moved to a better site. Salt water should not be used at potential electrodes because it causes excessive voltage drift.

The object in using porous-pot electrodes is to prevent development of contact potentials and to minimize electrode polarization between the electronically conducting metal contact electrode and the ionically conducting ground.

The porous-pot electrode consists of an unglazed ceramic pot or a plastic vessel with a porous or unglazed ceramic bottom part. The pot is filled with an electrolyte, usually copper sulfate, which leaks through the unglazed ceramic in contact with the moist ground. The cover of the pot holds a copper electrode in contact with the copper sulfate.

In theory, there is no electrochemical potential generated at a non-polarizable electrode, but in fact such an electrode is difficult to make, even in the laboratory. The porous-pot electrode is actually a 'buffered electrode' at which some potential drop is created if current is passed. This potential is small if proper care is taken to clean the pot inside and out and to use a saturated copper sulfate solution. To keep the unglazed ceramic free of calcium carbonate deposits use distilled water to make up the copper sulfate solution whenever possible and periodically soak the pot in hydrochloric acid.

Porous-pot electrodes are not always necessary if the voltmeter-receiver is ac coupled, because the very low frequency electrode potentials are effectively filtered out. However, in this situation, received voltages tend to be 'noisy' and true self-potential voltages are not obtained, which can be a disadvantage. Variation of resistance at contacts creates the unwanted noise. Some dc coupled voltmeter-receivers, such as the Newmont type, have automatic SP buckout capabilities, in which case use of porous pots can sometimes be avoided.

If not a porous-pot, the potential electrode is usually a sharp thin steel pin that can be thrust into the moistened earth. Other types of potential electrode contact materials can be used, depending on the field situation.

Reading the receiver

To conserve power, the receiver is switched 'off' between readings and the solid-state circuits need no warm-up time. The reading sequence varies between different instruments, so the following is only a generalization. After turning the voltmeter 'on' the noise-level is checked and the self-potential controls adjusted. Then a transmitted signal is called for after the voltage selector is set on the appropriate scale. Usually the strongest voltage

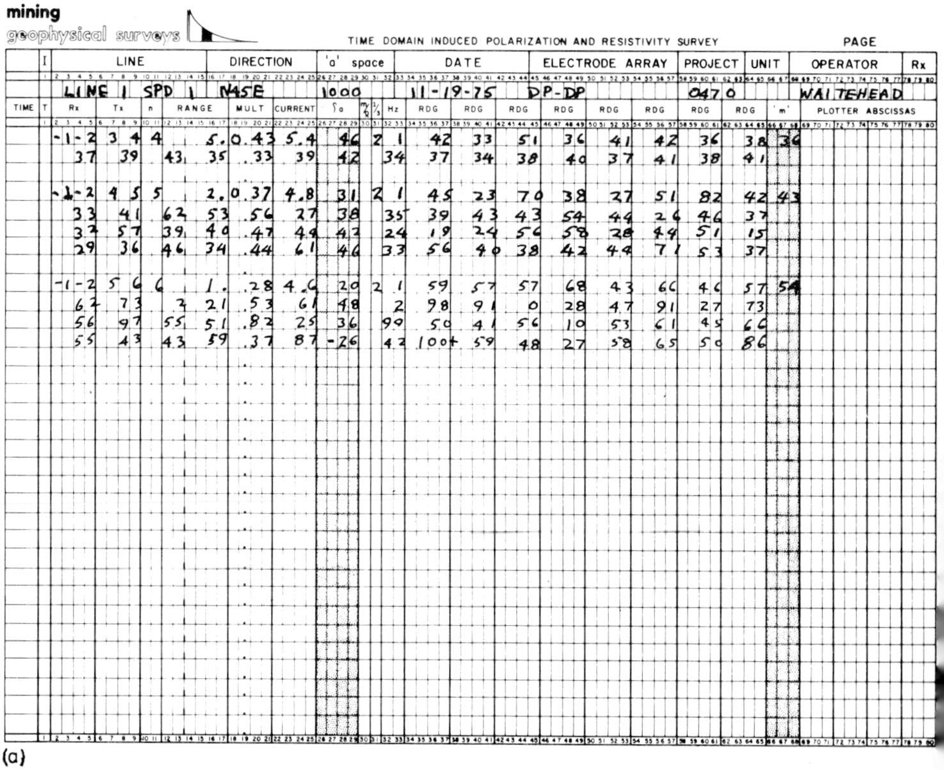

mining geophysical surveys — TIME DOMAIN INDUCED POLARIZATION AND RESISTIVITY SURVEY — PAGE

I	LINE	DIRECTION	'a' space	DATE	ELECTRODE ARRAY	PROJECT	UNIT	OPERATOR	Rx
	LINE 1 SPD 1	N45E	1000	11-19-75	DP-DP	0470		WAITEHEAD	

TIME	T	Rx	Tx	n	RANGE	MULT	CURRENT	ρ_a	m/R	1/3	Hz	RDG	RDG	RDG	RDG	RDG	RDG	RDG	RDG	'm'	PLOTTER ABSCISSAS
		-1-2	3 4	4	5.	0.43	5.4	46	2		1	42	33	51	36	41	42	36	38	36	
		37	39	43	35	.33	39	42			34	37	34	38	40	37	41	38	41		
		-1-2	4 5	5	2.	0.37	4.8	31	2		1	45	23	70	38	27	51	82	42	43	
		33	41	62	53	.56	27	38			35	39	43	43	54	44	26	46	37		
		32	57	39	40	.47	44	42			24	19	24	56	58	28	44	51	15		
		29	36	46	34	.44	61	46			33	56	40	38	42	44	71	53	37		
		-1-2	5 6	6	1.	.28	4.6	20	2		1	59	57	57	68	43	66	46	57	54	
		62	73	2	21	.53	61	48			2	98	91	0	28	47	91	27	73		
		56	97	55	51	.82	25	36			99	50	41	56	10	53	61	45	66		
		55	43	43	59	.37	87	-26			42	100+	59	48	27	58	65	50	86		

(a)

Fig.7.6a. Recorded field data.

signals are read first in order to allow the pot potentials to come to equilibrium. If noise is present at the lower voltage levels, additional time must be taken to get a good reading. All voltage values are recorded and reread to double check and avoid reading errors. After a receiver reading sequence is completed, the transmitter is called to be shut down and the pot man told to advance to the next station.

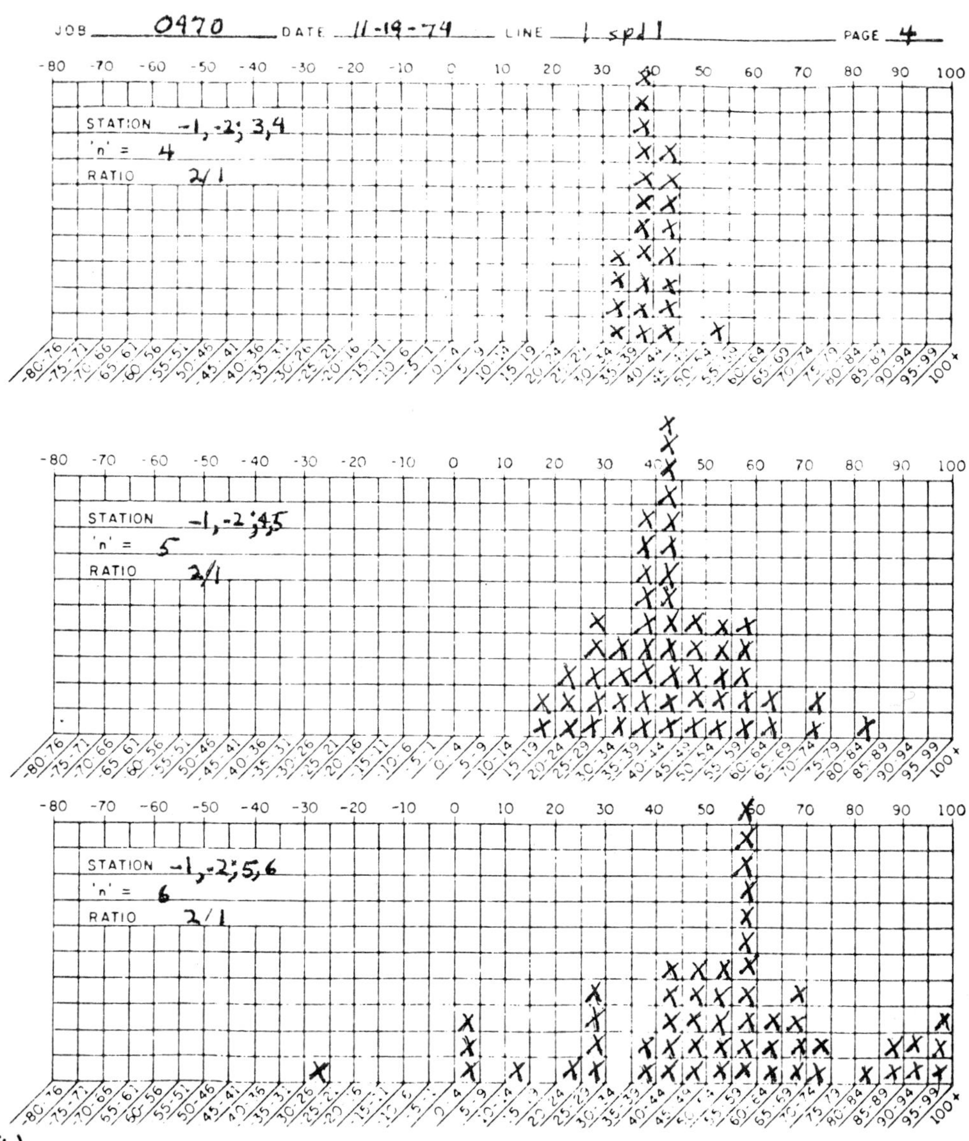

Fig.7.6b. Histograms.

Recorded field data (Fig.7.6) are reduced to the conventional polarization parameters and apparent resistivity, and these values are plotted graphically for interpretation. Large polarization anomalies can be recognized in the field, but more subtle patterns require mapping, correlation with other data, and perhaps a quantitative interpretation. If possible, it is always good field practice to continue a traverse past an anomalous zone, so that the limits of the feature can be determined. Zonge (1972) has described the interesting use of a minicomputer in the field for making IP phase spectral measurements over the frequency range of 0.1—100 Hz.

IP SURVEY CONDITIONS

After the objectives and coverage of an IP survey are decided upon, a decision must be made concerning the desired resolution of a subsurface body and the depth of exploration of the survey. It is not possible to have both high resolution (recognizing small features) and a great depth of exploration. The limitation on resolution of a subsurface body is its depth of burial: the scaling factors of electrical surveying allow a large deep body to be detected as readily as a small shallow body.

The limitations on depth of exploration, meaning the depth at which a given body can be detected as a distinctly anomalous feature, are dictated by the electrode spacing between the electrodes of a pair, the amount of signal that can be generated and detected, and the amount of interfering noise through which the signal must be distinctly recognized. The electrical properties of the earth impose the subsurface condition under which the electrical measurements are made.

After taking these important factors into account, together with resolution, depth of exploration, surface environmental conditions, and equipment characteristics, one gains an appreciation for the complexity in laying out an optimum IP survey. At best we can only approach maximum efficiency conditions as a goal in designing IP survey techniques.

Signal-to-noise ratio

It is possible to determine the signal-to-noise ratio for an IP survey layout, and this provides information as to the capabilities of the system. In fact, signal-to-noise ratio is probably the most diagnostic parameter in evaluating a field IP system. Unfortunately, there is no easy way to evaluate the amount of noise that passes through the IP voltmeter-receiver's filter. However, logically assuming that noise rejection is a constant fraction of the amount of noise received, conditions and even the effectiveness of different arrays can be compared on a relative basis.

The amount of voltage signal of an array is explicitly given by the apparent resistivity formula for the array. For example, for the dipole–dipole array, the signal voltage V_s from the transmitter is given by the expression developed in Chapter 2:

$$V_s = \frac{\rho I}{\pi a(n+1)(n+2)n} \tag{91}$$

where ρ is resistivity, I is transmitted current, a is the electrode spacing, and n is the integral number of electrode spacings a between the adjacent transmitter and receiver electrodes, as shown in Fig.2.10.

The amount of noise can be extremely variable but is best expressed in eq. 80, Chapter 6, and can be rewritten for comparative purposes as:

$$V_n = Ka\sqrt{\rho} \tag{92}$$

The filtering constant K is mainly dependent on the efficiency of the receiver and the noise-rejection method; a is the receiver array interval and ρ is the apparent resistivity of the subsurface.

The ratio of signal to noise is the ratio of eq. 91 to eq. 92 and for the collinear dipole–dipole array is:

$$\frac{V_s}{V_n} = \frac{\rho^{1/2} I}{K \pi a^2 (n+1)(n+2)n} \tag{93}$$

Equation 93 is a general expression for signal-to-noise ratio, assuming that a constant fraction of noise is rejected by a given receiver. The threshold ratio is plotted on Fig.7.7 for different values of resistivity at a fixed electrode

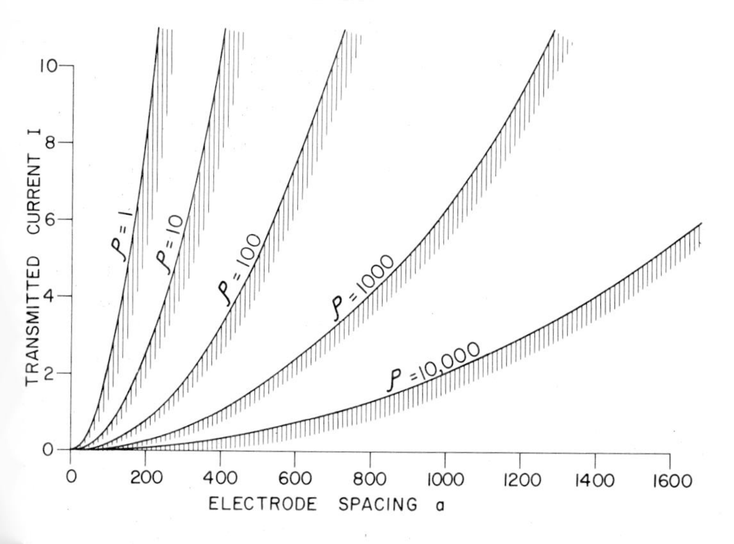

Fig.7.7. Resistivity values plotted at a constant signal-to-noise threshold as functions of transmitter current I and electrode separation a at a distance of $5a$ between dipoles.

distance of $5a$. This figure shows how the signal-to-noise ratio diminishes in low-resistivity ground. Figure 7.8 illustrates how the signal-to-noise ratio decreases with n assuming a fixed value for the resistivity.

Comparison of electrode arrays in IP surveying

Although there are a large number of variables in the choice of arrays for IP surveying, a relative comparison can be made. This is given in Table VIII. In order of relative general importance, the factors being weighted are:

(1) Signal-to-noise ratio.
(2) EM coupling rejection.
(3) Survey speed and economy.
(4) Resolution of subsurface bodies.
(5) Array symmetry.
(6) Other matters, such as safety, topographic effect, communications, ease of interpretation, and so forth.

It is obvious that some arrays are better suited for certain geological conditions or for certain search problems. For example, for large array intervals where wire length and its resistance become a problem, the collinear dipole—dipole array becomes more suitable.

Basically, the advantages of the time-domain IP method are that EM coupling is more simply and effectively rejected and the entire IP response is ideally recorded in a single decay curve. But telluric noise is often a problem with the pulse IP method.

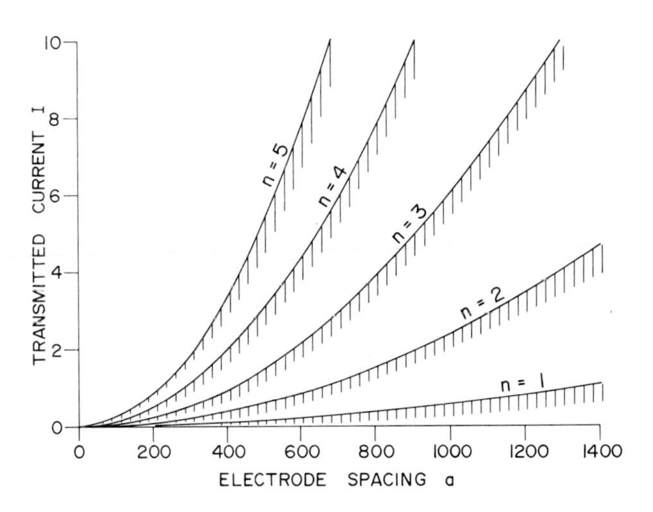

Fig.7.8. Constant signal-to-noise threshold shown with electrode spacing multiple n plotted against transmitter current I and electrode separation a at a resistivity of 10 Ω m.

TABLE VIII

Comparison of IP survey electrode arrays

Arrays	Advantages	Disadvantages	Survey speed	*S/N*	EM coupling rejection
Parallel field arrays					
Wenner	Anomalies symmetrical Synchronous detector possible Many case histories available	Requires more wire, larger field crew Poor resolution Unfavorable in capacitive coupling situations	Fair	Good	Fair
Schlumberger	Symmetrical array Synchronous detection possible Fewer men required Works well in layered earth Type curves available	Less horizontal resolution Unsuitable for horizontal profiling	Fair	Fair	Fair
Gradient	Map interpretation easier Less masking by conductive overburden Penetration good Safer Communications easier Can use two or more receivers Less topographic effect Data easily contoured in plan Useful where difficulty in making good current contacts	Poor resolution with depth Poor in low-resistivity areas Geometric factor varies complexly	Good	Fair	Poor

Continued overleaf

TABLE VIII (continued)

Arrays	Advantages	Disadvantages	Survey speed	S/N	EM coupling rejection
Potential-about-a-point					
Three array	Good reconnaissance array Fairly good resolution	Asymmetrical More wire needed	Fair	Good	Good
Pole—dipole, collinear	Good resolution Good subsurface coverage	Asymmetrical	Fair	Fair	Fair
Perpendicular three array, pole—dipole, pole—pole	Virtually eliminates EM coupling	Asymmetrical More wire required	Fair to good	Fair	Very good
Pole—pole (two array)	Smaller crew needed Less wire needed than for some arrays Good penetration in non-conductive overburden	Susceptible to masking by conductive overburden	Good	Poor	Poor
Potential-drop ratio (PDR)	Sensitive to lateral variations 'Common mode' noise rejection	Complex interpretation	Fair	Good	Fair
Dipole field array					
Dipole—dipole, collinear	Symmetrical, good resolution Good penetration Less survey wire needed	Slow unless equipment is portable Resistivity topographic effects Interpretation somewhat involved	Fair	Poor	Good
Dipole—dipole,	Special use for EM	Not used for routine	Poor	Poor	Good

TABLE VIII (continued)

Arrays	Advantages	Disadvantages	Survey speed	S/N	EM coupling rejection
Downhole arrays					
Azimuthal array (one potential electrode down hole)	Fair for exploration Useful in finding best search direction Useful in obtaining rock properties	Interpretation complex Negative anomalies	Fair	Good	Good
Radial array (one current electrode down hole, mise à la masse)	Good for exploration Useful in finding best search direction Hole need not stay open	Interpretation complex Negative anomalies Not good for obtaining rock properties	Fair	Good	Good
In-hole arrays					
More than one electrode in hole	Good for obtaining rock properties Good for assaying Interpretation simple	Current densities may be too large Possible capacitive coupling problems Not designed for exploration Special equipment Expensive	Good	Fair	Good

Frequency IP equipment can more effectively filter out the telluric noise, but EM coupling is a problem. The IP spectral method gives the most information about the electrical properties of the subsurface, but the method is slower and more costly than the conventional method.

Safety

Insurance rates indicate that geophysical surveying is hazardous. Yet upon study of the situations that lead to accidents, the human factor predominates rather than equipment malfunction. As is true in other human endeavors, accidents can be avoided by using careful procedures and by keeping equipment in good condition.

Vehicular accidents are a leading cause of injury and equipment damage. Trucks and cars used in IP surveying are often driven in remote areas under hazardous conditions. Little can be done to prevent these accidents besides providing safety features in the vehicles and applying the rules of safe driving.

The field environment of IP surveys is often hazardous to the unwary person. Unfriendly animals, reptiles, and insects are frequently plentiful. The desert vegetation is thorny, the jungle is steamy, and high altitudes and latitudes are cold. Men and equipment will not operate efficiently and properly unless they are adapted to environmental conditions.

A well-stocked medical first-aid kit and instruction book should be provided to each crew vehicle. The contents may vary, depending on whether venomous snakes are prevalent or axes are being used. Each vehicle should have an approved-type fire extinguisher and a separate CO_2 or powder-type extinguisher should be available for the motor-generator and transmitter.

Electrical currents are a primary hazard in IP field surveying. Most of the electrical burn and shock accidents stem from communication difficulties between the transmitter operator and a field man who is attempting to establish a current electrode or lower its resistance. Severe hand burns can result when a 'hot' wire is brought up to the electrode. Shock and even death are possible, particularly if the victim's body forms part of the current path. Two-handed splices should be avoided at a current electrode.

The effects of electricity on the human body have been described by Molinski (1970). It is interesting to note that low-voltage ac circuits are more dangerous than corresponding dc circuits, but that contact with high-voltage ac circuits is less likely to be fatal than contact with corresponding dc circuits. The value of the alternating current, which is used in IP surveying, that flows through the body when contact is made is of extreme importance in determining the resulting injury. Currents above 8 mA are considered unsafe. Below 8 mA, the sensation of shock is not painful, while between 8 and 15 mA, the shock becomes painful but muscle control is not lost and the victim can let go. Between 15 and 20 mA, control of adjacent muscles is lost and the victim cannot let go, and between 20 and 50 mA severe muscular contractions occur and breathing becomes difficult. Currents between 50 and 200 mA cause a potentially fatal condition, ventricular fibrillation, in which the heart muscles lose rhythm and coordination and the twitching heart can no longer pump blood. Ventricular fibrillation requires the application of a counter shock to bring the heart muscles to rest and permit return to a normal rhythm. Above 200 mA, a block or partial paralysis of the nervous system occurs which results in cessation of normal breathing and in severe muscular contractions that stop the heart action by clamping. The paralysis of the nervous system may last for several minutes or hours before normal function returns, and the victim requires immediate mouth-to-mouth resuscitation which must be continued until recovery or medical help is available. In general, low voltages kill through the mechanism of ventricular fibrillation and high voltages either through destruction or inhibition of nerve centers, asphyxia being the immediate cause of death.

When respiration ceases and the heart stops, the victim may appear to have 'died'. Indeed, he will die unless artificial respiration and cardiac resuscitation are administered and are continued until there is either recovery or true death. Medical treatment may be necessary to restore heart action, and artificial respiration or mouth-to-mouth resuscitation should be continued until medical help is available.

Thunderstorms near electrical survey equipment and crews are a dangerous situation. Work should be abandoned and electrode connecting wires removed from the terminals of the transmitter and receiver when lightning clouds approach. The interior of a metal-enclosed vehicle is a moderately safe place during a severe thunderstorm; otherwise, lie on the ground in an open area away from grounded objects.

Chapter 8

DRILL-HOLE AND UNDERGROUND SURVEYING AND THE NEGATIVE IP EFFECT

In some ways a subsurface IP investigation is simply an extension of surface surveying. The electrode arrays used are similar to those for resistivity and IP surveying described in Chapter 2. Equipment and techniques are also similar to those for surface surveying, with a few notable exceptions. The negative IP effect is encountered in many survey situations and is especially seen in drill-hole and underground IP surveys, so it will also be discussed in this chapter. But some important differences between surface and subsurface IP methods exist and these will be discussed under the following three headings.

DIFFERENCES BETWEEN SURFACE AND SUBSURFACE METHODS

Theory

Geometric factors used for obtaining apparent resistivity in subsurface surveying differ by two from surface geometric factors. For example, the apparent resistivity for a Wenner array (Chapter 2) used on the surface is $2\pi a(V/I)$, while if used in the subsurface it is $4\pi a(V/I)$. This is because the electric field of an entire sphere must be taken into account rather than that of a hemisphere, as was shown in eq. 16. If the electrode array is only partly on the surface and partly underground, then an intermediate geometric factor will apply. It is not easy to derive the geometric factor for an electrode array that is partly on surface and partly down a hole (usually termed a 'downhole array'). Also, the presence of drill casing may complicate the problem.

The IP effect, $\Delta\rho/\rho$ does not change with burial of the electrodes because induced polarization is measured as a ratio of potentials with respect to changing time or frequency. However, unusual IP effects are frequently found when electrodes are very close to sulfide bodies, because of the directionality of polarization.

Survey method

Electrode arrays used for subsurface surveying are restricted to those that can be readily adapted to the available fixed openings in the subsurface. This restriction of electrode mobility is in considerable contrast to surface surveying, where electrode positions can usually be established wherever desired.

In subsurface work, the direction in which polarization information is sought may vary from the conventional sense of a surface survey. For example, it may be the objective of a particular subsurface IP survey to determine the rock properties between electrodes, either for assay-type information or for making a mineral discovery. On a surface survey, the surface material between electrodes may be readily examined, whereas in subsurface work this area is inaccessible.

The electrical environment differs considerably between the surface and the subsurface, requiring a modification in exploration approach in going from one region to the other. High humidity, temperature, and pressure conditions can usually be anticipated. Therefore, survey methods, equipment, and electrodes used for subsurface work can differ from those used on surface projects.

Interpretation

The interpretation of subsurface data is generally more complex than that for surface-derived electrical survey information. For example, a simple layered geometry for which there are master curve solutions seldom exists in subsurface surveying. Conductive subsurface bodies can sometimes be viewed as having symmetrically located virtual images, and this causes an interpretational ambiguity when dealing with subsurface electrical data. This uncertainty can sometimes be removed by using different electrode intervals or arrays.

Metallic polarizable minerals can lie very near surveying electrodes or between them when using a subsurface array. This situation would be unusual for a surface array because of near-surface oxidation of metallic minerals and the soil cover ordinarily present. When polarizable minerals lie in the plane of the array or near the surveying electrodes, peculiar IP effects are encountered, particularly negative IP effects, which can complicate the interpretation of subsurface IP data.

The mise-à-la-masse method whereby a downhole current electrode is employed to create a potential field pattern around a conductive body is an effective means of drill-hole resistivity surveying. Certainly, the resistivity results can indicate the direction toward better mineralization. However, if a current electrode is emplanted in a mineralized body for the purpose of doing a drill-hole IP survey, peculiar data can result. This is caused by two factors. The high current densities near the electrode result in a nonlinear IP response. Also, if the current source is within a polarizable body, a monopolar rather than a dipolar IP effect is produced, as was indicated by eq. 52 of Chapter 3.

TECHNIQUES FOR SUBSURFACE IP AND RESISTIVITY SURVEYS

From the standpoint of survey objectives, there are two principal categories of subsurface IP and resistivity surveys. One is the examination of the electrical properties of the rock lying between and near the electrodes, as in a close-spaced electrode interval drill-hole survey, while the other is the discovery of a mineral deposit or the determination of the direction toward better mineralization. These differences in objectives for subsurface surveying bring about a natural division between the methods and equipment used and between the interpretation of data.

Hole logging for rock properties

A drill hole offers an already prepared surface with regular geometry from which precise measurements of in-place electrical properties can be made for different purposes. Figure 2.20 is an illustration of an electric log of a drill hole. The measurement process is accomplished by putting two or more electrodes in the hole at predetermined positions and taking resistivity and IP values. These values plotted against hole depth constitute a 'hole log' of electrical properties that correspond closely to the mineralogic and other properties of the rock. If the electrodes are further apart, the electric log represents an average of the properties of the rock within a greater distance from the hole. The purpose of an in-hole electric logging may be assaying, a study of the properties of in-place rock, or correlation of electrical data from one drill hole with that of another.

There are several types of electrode arrays and intervals (Fig.8.1) that can be used for electric well logging, depending on the purpose of the survey. The limitations of resolution and depth of exploration apply in this method in the same way as in surface surveying.

Assaying

IP drill-hole logging has been used to 'assay' the amount of metallic material in the proximity of the drill hole. This method can really only be applied safely to give consistent results in situations where it has been established that there is a simple, direct relationship between the IP effect or metal factor and the metallic mineral content.

An advantage of the proper use of the IP assay method is that lower cost percussion drilling methods can be used instead of core drilling. Also, the electrical assay array can sample a large volume of rock. This technique was successfully employed by Schillinger (1964) and Bacon (1965) in exploring the native copper deposits of the Upper Peninsula of Michigan. Here the IP assay method was successful in determining whether the copper-bearing rock near the hole was of ore grade and thus allowed selective mining of the deposit. The three-array was used in this project with both pulse and

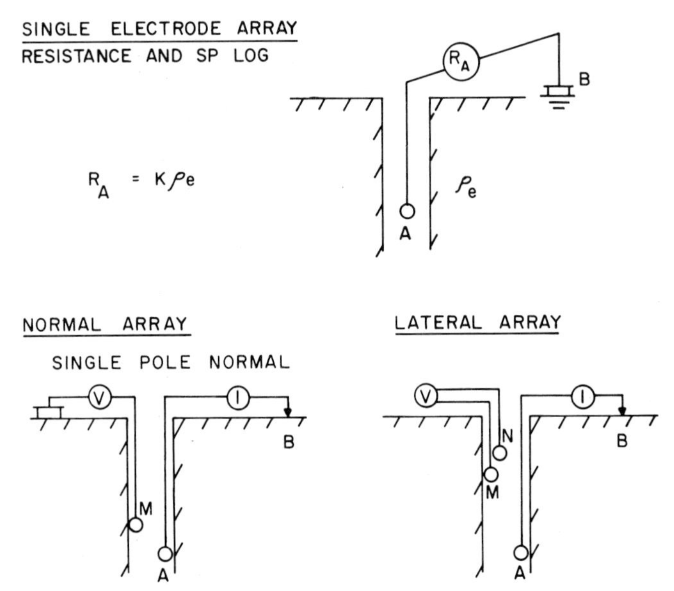

Fig.8.1. In-hole logging arrays.

frequency methods. Chargeability or frequency effect and metal factor were used to estimate native copper values. When negative or questionable IP readings were obtained, more reliance was placed on the resistivity readings.

Before a given rock type can be assayed electrically, studies must be made of the electrical characteristics both of the material of interest and of all the rock types in the vicinity. A statistical analysis correlating metallic mineral content with IP response and metal factor must be made so that the relative reliability of the IP assay method is known. In addition, a cost analysis may be carried out to learn the extent of savings.

In general, there is no simple direct correlation between IP or metal-factor response and metallic mineral content of a rock. Correlation is entirely dependent on the pore fabric of the rock, the textural habits of the metallic minerals, and the interrelationship between the minerals. Komarov (1970) and Maillot and Sumner (1966) illustrate correlation relationships in some specific situations.

Rock properties

There is a strong argument for the continued use of in-hole drill-hole logging for studying the in-place IP properties of rocks. In contrast to laboratory sample measurements, interstitial fluids are not disturbed and the characteristics of the rock itself can be carefully and perhaps extensively studied by analysis of the drill core.

The prepared surface of the drill hole is usually smooth, but irregularities can cause trouble. The presence of drilling fluid influences measurements, and the drilling fluid also penetrates cracks in the wall of the hole. In addition, disturbing directional resistivity and polarization effects may take place near the measuring electrodes, and current densities above 0.1 $\mu A/cm^2$ may cause a nonlinearity of the IP response, particularly at lower frequencies or longer charge times.

Correlation of electrical data

In-hole IP surveys are being carried out by mining and exploration companies as a means of correlating beds and distinctive rock types from one hole to the next by electrical characteristics or for general geologic information. Electric logging has been extensively used for correlation in petroleum exploration because drill core is not commonly available. Any correlation of electrical properties with rock type is often only a local relationship, particularly in 'hard-rock' mining situations, so caution is necessary in applications of this method. However, drill-hole correlation methods are relatively straightforward and reliable and have a good record of effectiveness in ore finding (Salt, 1966).

Exploration drill-hole surveying

To expand the effective radius of search of a drill hole, a single electrode can be put down the hole and the surrounding area surveyed to determine if there are anomalies in the overall electrical field pattern observed. Such a survey procedure also uses the drill hole to search deeper into the subsurface than is possible if only surface electrode locations are used. This method would presumably allow a savings in drilling costs. Either a current or a potential electrode can be put into the drill hole.

By the theorem of reciprocity (Chapter 2), an exchange of current and potential electrodes would cause no difference in interpreted electrical characteristics. However, current electrodes are more difficult to establish than potential electrodes, thus an array would usually be chosen in which the potential measuring electrodes are the mobile ones.

Mise-à-la-masse and azimuth direction to mineralization

One of the simplest yet most effective downhole geophysical survey methods is to energize a mineralized zone by emplanting a current electrode within it or near it, then to measure the resulting potential field distribution. This field distortion is illustrated in Fig.2.14a, which shows the deflection of equipotential surfaces near a conductive body. The method works particularly well if a drill hole has encountered mineralization and it becomes necessary to find the trend of the mineralized zone in order to plan additional drill holes (Fig.8.2).

The distribution of potentials or the shape of the equipotential surfaces is indicative of the geometric trend of the subsurface conductor and it may

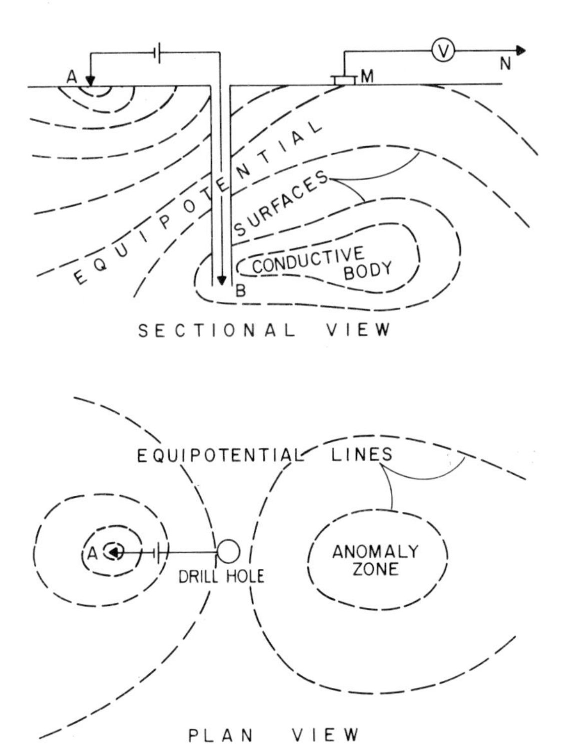

Fig.8.2. The mise-à-la-masse method.

give information about the shape of the body, its extent, dip, and plunge. In simple situations, the voltage potentials can be measured at the surface, but they have been successfully mapped in drill holes or in mine openings. Morrison (1970) illustrates the use of the 'direction to better mineralization' method in the lead—zinc belt of southeastern Missouri, and McMurry and Hoagland (1956) made good use of in-hole potentials in outlining mineralization at Austinville, Virginia. The mise-à-la-masse method is highly developed in Europe, and Parasnis (1973) has a good summary of the method with illustrations of successful applications.

A note of caution must be made concerning problems in interpreting the meaning of IP data gathered while using the mise-à-la-masse method. Difficulties arise because resistivity contrasts near the polarizable body can distort the polarizing and depolarizing currents in perplexing ways, requiring a very careful analysis of the meaning of the IP data. In an anisotropic rock environment, an apparent anomaly measured on the earth's surface may actually be due to the peculiar focusing nature of subsurface electrical anisotropism. This not uncommon rock property also may shift a surface electrical anomaly a considerable lateral distance from the cause of its source.

Downhole survey methods

Downhole survey methods use a drill hole to extend knowledge of the electrical properties of the subsurface. Basically, there are two ways of doing this type of surveying, either the use of a current electrode down the hole with the potential electrodes on the surface or the use of a potential electrode down the hole with the current electrodes on the surface. The arrays are illustrated in Fig.8.3. Current electrodes are not as easily emplanted as are potential electrodes, and Fig.8.3a shows the more mobile potential electrode pair being used to log electric fields in the drill hole that are due to surface current electrodes that straddle the hole. The current polarity is significant in the interpretation of the location of a possible polarizable body. For example, the polarization decay waveform will be either positive or negative, depending on which side of the hole the polarizable body is located. By using pairs of current electrodes at different azimuths across the hole, it is possible to interpret the direction to a nearby polarizable body and perhaps its shape. This method has been used by Newmont Exploration Limited using time-domain equipment, and it has been described by Wagg and Seigel (1963) and Dolan (1973). In the azimuth method of downhole surveying, the surface current electrodes are usually placed at a distance from the drill hole equal to three-quarters of the depth of the hole.

The downhole radial method consists of a system of three current electrodes as shown in Fig.8.4. There is one current electrode (D) down the drill

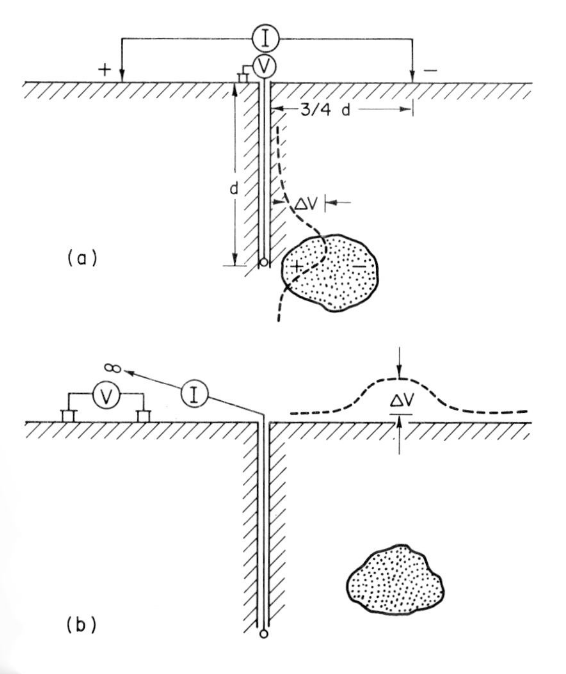

Fig.8.3. Downhole electrode arrays: (a) azimuthal IP method and (b) radial IP method.

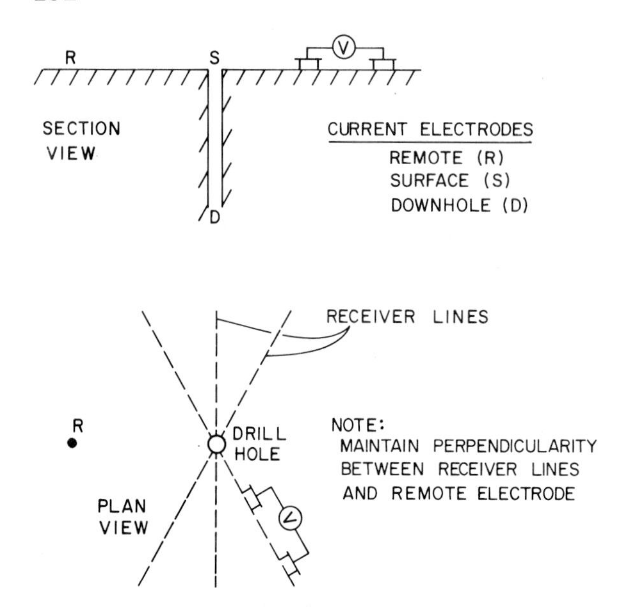

Fig. 8.4. Downhole radial IP method using three current electrodes.

hole, one (S) at the surface near the collar of the hole, and a remote electrode (R) on the surface several hundred meters from the hole. Current is sent between three different pairs of electrodes, S—R, D—R, and S—D. Voltage and IP measurements are made on the surface along radial lines extending from the drill hole. These lines are approximately at right angles to the remote electrode so as to minimize electromagnetic inductive coupling. With the three-current electrode system, the impedance measured between any two points on the surface can be checked by using different transmitter dipoles to give a measure of data quality and accuracy. Hauck (1970) has discussed this system in detail. Peterson and Sumner (1973) give geometric factors and show how to correct for the presence of near-surface casing. Figure 8.5 displays contoured results of the radial downhole method, and Figs. 8.6 and 8.7 illustrate the desirability of having the current electrode as deep in the hole as possible in order to explore for deep ore bodies.

Underground IP surveying

IP surveys can be carried out in underground mine openings for assaying or ore finding. However, the underground environment is hard on equipment, and the survey must be designed to trace mineralization lying between the mine workings or beyond them. Thus, underground IP surveys are specialized in their nature and purpose, and they are usually used to supplement other exploration methods.

Sumi (1961) found that underground IP data were helpful in nonmetallic underground exploration in Yugoslavia, and Mathisrud and Sumner (1967)

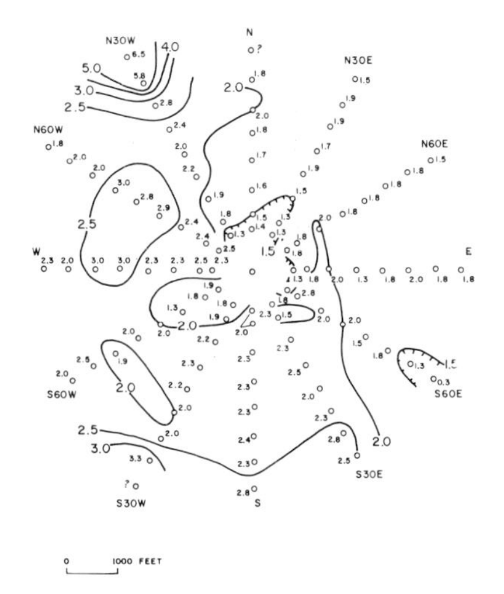

Fig.8.5. Contoured results of the radial downhole method, showing percent frequency effect using the downhole-to-remote electrodes. After Peterson and Sumner (1973).

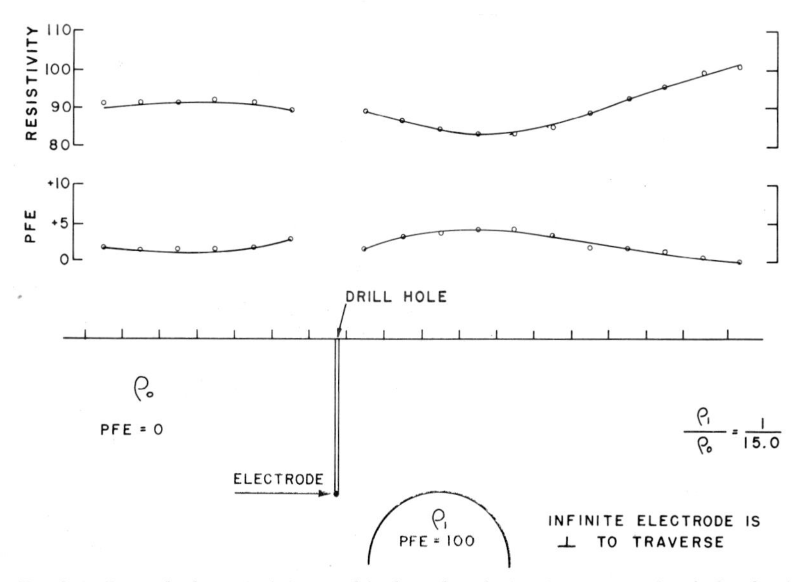

Fig.8.6. Downhole resistivity and induced polarization near a buried spherical body.

attribute discoveries at the Homestake mine in Lead, South Dakota, to the method.

Interpretation of underground IP data is complicated by the existence of conductive track, pipe, and mine water in drifts. Electrode arrays must be chosen with care because of possible negative effects. For example, a polarizable body lying between electrode pairs of the dipole—dipole array will give rise to negative results (Fig.8.8), but the gradient array is susceptible to capacitive and electromagnetic inductive coupling.

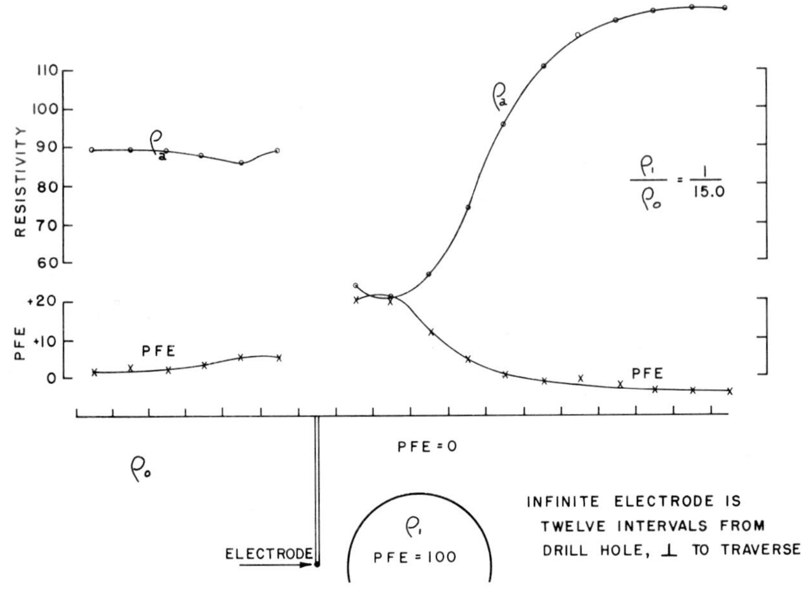

Fig.8.7. Downhole resistivity and induced polarization near a spherical body buried shallower than that of Fig.8.6.

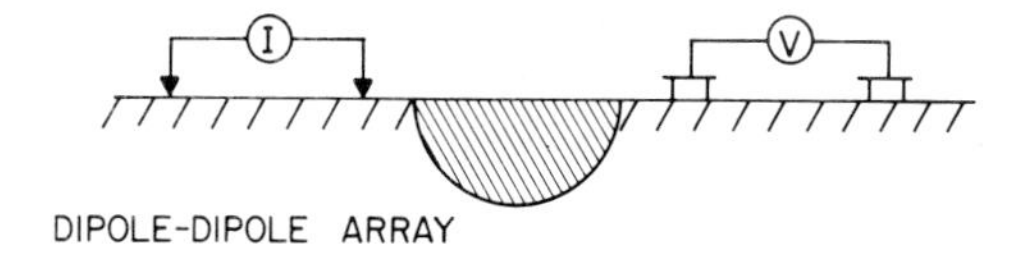

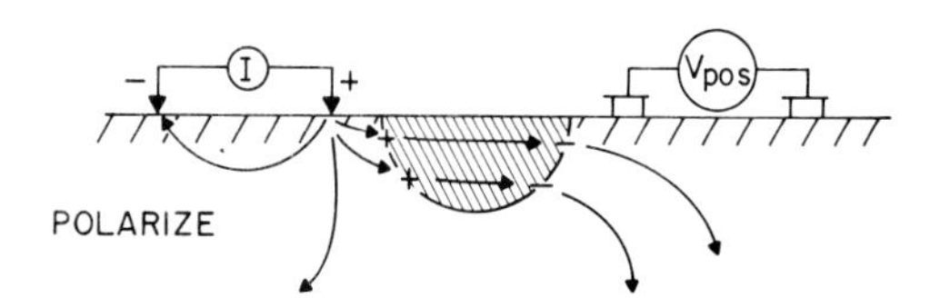

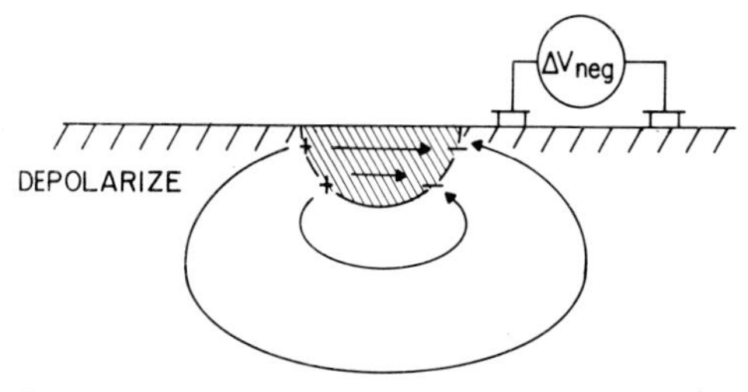

Fig.8.8 Negative IP effects due to a near-surface body.

NEGATIVE IP EFFECTS

Negative IP response is due to the vectorial electrical field relationships of a polarizable body during the polarizing and depolarizing cycles. Figure 8.9 illustrates the conventional IP response to be expected at ground surface and the negative response that would be observable off the ends of a polarizable body. In fact, because of the character of a dipolar field, negative IP effects can be even larger than positive effects.

To understand the reason for the negative IP effect it is convenient to imagine surface charges forming on a polarizable body during the charge cycle as was discussed on p. 70. During the discharge cycle these charges will flow back to establish electric neutrality and if the voltage-detecting dipole is oriented to detect a reversal in flow direction, this is seen as a negative response. A highly conductive polarizable body will internally short out this depolarizing current. The resulting considerably diminished observable IP response supports the argument for the metal factor's being a valid IP measurement parameter. The presence of material with resistivity contrasts surrounding a polarizable region can also create an unusual IP response pattern.

Negative IP effects are not unusual adjacent to bodies of limited lateral extent, especially if the bodies are relatively near the electrode contact points.

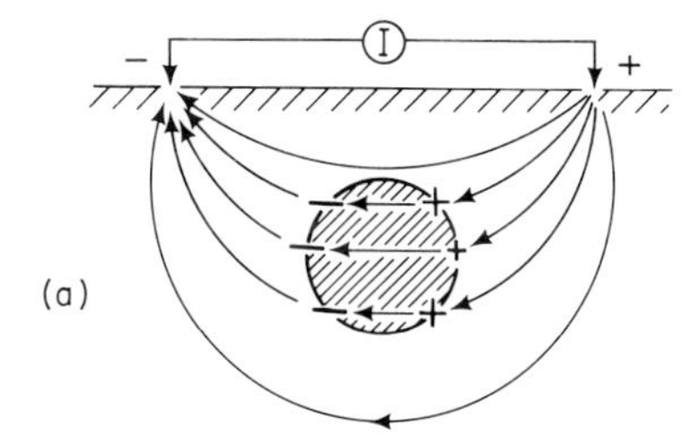

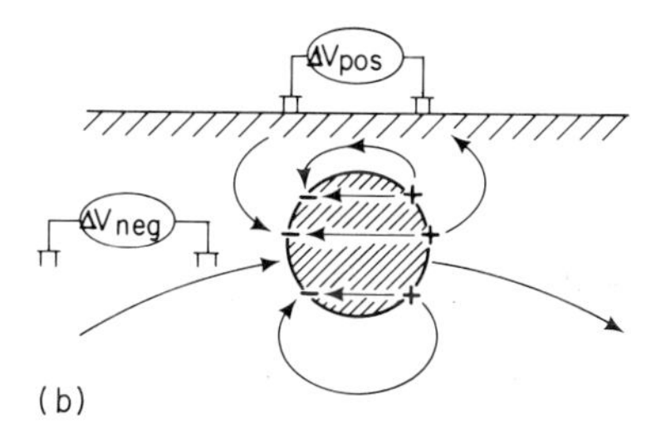

Fig.8.9. Negative IP effects due to a buried body: (a) the polarization cycle and (b) the depolarization cycle.

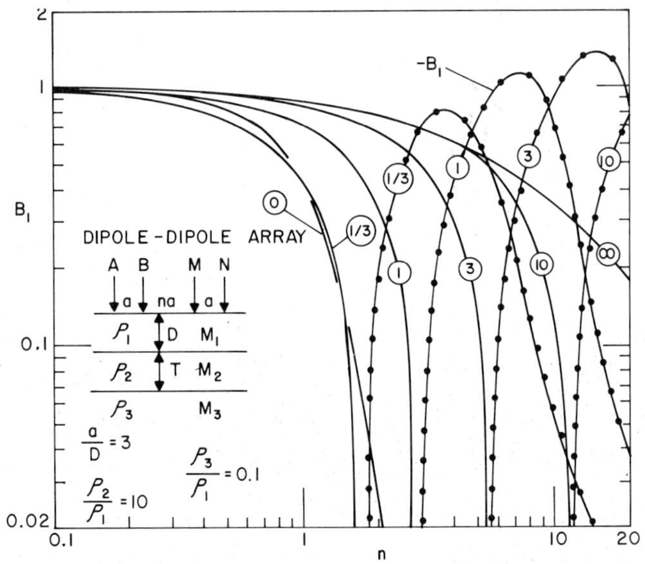

Fig.8.10. Three-layer IP curves showing negative effects. After Nabighian and Elliot (1974).

However, Nabighian and Elliot (1974) have pointed out that negative IP effects can be found in horizontally multilayered situations where the lowest layer is more conductive than the layer immediately above and the uppermost layer is at least somewhat polarizable. A family of IP response curves for surveys over three horizontal layers based on eq. 61 is shown on Fig.8.10.

Negative IP values must be interpreted with care. Positive IP values are usually associated with negative IP values in definite patterns. This is because induced polarization is bipolar in the depolarization process. For most surveys, positive values are larger than negative. If these conditions are not seen in the data, there may well be equipment or coupling problems.

Chapter 9

INTERPRETATION OF INDUCED-POLARIZATION DATA

The induced-polarization method has become a routine, accepted geophysical method in exploring for concealed mineral deposits. There is a wide variety of IP techniques that use different kinds of instruments and equipment to accomplish the field work. The more sophisticated systems encourage the acquisition of large quantities of data at a single location and the use of a portable computer to process much of the data in the field. This chapter will discuss the interpretation of IP data and the total interpretation of a geoelectrical survey.

The interpretation of IP field data actually starts with the original planning of the survey. Surveys are laid out with specific purposes in mind, whether general reconnaissance information is being sought or specific targets are hypothesized. It is helpful to the final interpretation of an exploration venture to know from statements written beforehand what the survey was originally intended to accomplish.

A properly planned survey takes into account the field conditions, equipment limitations, and the probable nature of the exploration target. With these conditions in mind, the survey planner can lay out the optimum traverse directions, line spacing, electrode interval, and bounding limits of the survey to accomplish its object. The data from a properly planned survey are much easier to understand and interpret than data taken in a random fashion.

Exploration managers are sometimes unaware of the capabilities and limitations of geophysical exploration methods because they were originally trained as mining engineers, geologists, or attorneys. Regardless of any management bias, the geophysicist must do the best professional job possible and be a strong cooperative member of the exploration team, while appreciating other non-geophysical aspects of the total exploration problem.

The interpretation of geophysical data is a curious mixture of art, science, training, experience, and intuition (or luck). Geological knowledge plays an important role in most exploration situations, and the interpreter should have some geological background or at least have near at hand a geologist who knows the area. Geological reasonableness is a continual test of any geophysical model, and geological conditions in the form of body geometry and physical property contrast supply the boundary conditions for a geophysical interpretation.

The principle of synergism applies quite rigorously to geophysical interpretation. This is to say that the amount of information gained by several geophysical methods is greater than the sum of the individual interpretations. In this context, known geology, both from surface mapping and perhaps drilling, provides its own individual interpretation to be integrated into the whole. Here the geophysicist should be at the geologist's side in the geological interpretation, because the geologist frequently may not ask geophysical questions of his data.

Fortunately, the IP method provides several individual sets of data, and although they are all electrical in nature, the methods are distinct in their own right. Thus, individual interpretations of induced polarization, resistivity, self-potential, and electromagnetics can be made from a comprehensive electrical survey, and in some situations, a particular set of electrical data can also provide important subset information, as for example, both conventional IP and spectral IP data from the same survey.

METHODS OF INTERPRETATION

Case-history studies

The complete pragmatist may prefer to review the problem of IP interpretation mainly from a background of a comprehensive examination and analysis of exploration case-history examples. There is much to be said in favor of this approach to interpretation. Actual case histories obtained through personal field experience provide a valuable background to the interpreter, but very few people have been totally involved in a wide range of case-history activities. Also, poor case-history examples are often misunderstood, even by the people involved, and it is easy to attribute an exploration failure to an unfortunate, poorly understood experience.

Unfortunately, there really is not space in this volume to do total justice to an exhaustive case-history analysis of IP interpretation. The inquisitive reader is urged to review the many case-history examples of IP interpretation that are available in the literature, particularly those examples using similar equipment or in similar geological situations to the data being interpreted. *Mining Geophysics*, Vol. I (1967) and Vol. II (1969), published by the Society of Exploration Geophysicists, has several good papers and numerous references to worthwhile articles. The IP papers in *Geophysical Prospecting*, the journal of the European Association of Exploration Geophysicists, are good, and *Mining Engineering* (AIME) and the *Bulletin of the Canadian Institute of Mining and Metallurgical Engineers* have good case-history material.

Geophysical contractors have been generous in providing case-history information to prospective clients and professional associates. In particular, McPhar Geophysics Ltd., Scintrex Ltd., and Huntec Ltd. have a large variety of survey examples from over much of the world.

A final word of caution is necessary before applying the detailed results of an older case-history example to a given field problem being interpreted for evaluation. Case histories are often used to illustrate a particular exploration principle. But there is no assurance that identical conditions exist in the situation being interpreted.

While the IP method is most effective in the search for disseminated sulfide deposits, it has been proved useful for many other types of mineral occurrence. Measurement of rock properties is very helpful to the interpreter in the planning stage and in guiding the exploration to a successful conclusion. Some exploration managers may view a program as successful only if it has found ore, whereas a well-run exploration venture is in itself a noteworthy example of the efficient use of all exploration methods.

Physical model studies

The IP effect can be studied in miniature by simulation, using models. These models can be real polarizable bodies, but there is a problem in avoiding the too high levels of current density that may be necessary to obtain a strong polarization signal. Also, it has proved to be very difficult to create uniformly homogeneous materials with precise polarization properties and to control uniform resistivities and resistivity contrasts in model materials. Bertin and Loeb (1974) of CGG have described experiments using real polarizable materials on a small scale to simulate real field conditions.

Resistivity models, which are more easily controlled in their values than polarization models, can be used to simulate polarization effects by the theory described in Chapter 3. The idea of using two resistivity models to simulate polarization effects is simple but ingenious, and it was first used by Adler (1958) of MIT at the suggestion of Madden. Briefly, the analog model concept uses two resistivity models of identical size and shape but with different resistivity contrasts. Then two sets of resistivity data are taken using the same measuring electrode positions; the difference between the two sets of resistivity data is analogous to a polarization measurement. The measured difference in resistivities, both true and apparent, are proportional to the real and apparent polarization effects of a body of the same shape measured under the same geometric conditions.

If two identically shaped model bodies with resistivities ρ_1 and ρ_2 are used, the intrinsic, true analog model frequency effect would then be:

$$\mathrm{FE} = \frac{\rho_2 - \rho_1}{\rho_1}$$

where ρ_1 and ρ_2 are analogous to the high and low frequency responses, respectively. Then the apparent frequency effect $(\mathrm{FE})_a$ of a modeled situation would be:

$$(\mathrm{FE})_a = FE\,\frac{\rho}{\rho_a}\,\frac{\partial \rho_a}{\partial \rho} \tag{94}$$

where ρ is the intrinsic resistivity of the body. An analogous relationship holds for the time domain.

This principle of resistivity difference can also be applied to generate IP model curves from apparent resistivity curves. A large number of analog IP models were measured in an extensive research program carried out under the direction of P. G. Hallof by McPhar Geophysics Ltd. from 1966 through 1973. McPhar used a plastic model block impregnated with graphite and changed the resistivity of the surrounding alkaline solution to obtain a given set of data instead of using two model blocks.

Scale factor

A normalized set of apparent resistivity and IP values for a specific subsurface modeled structure can be linearly scaled to corresponding resistivities and dimensions to match an actual physical situation. This is an illustration of an important principle of similitude that applies in relating resistivity and IP scale models to field problems.

One interesting aspect of this principle applies to the question of how deep can the IP method explore. The answer lies in noting that the ratio of electrode interval to depth and body geometry can be scaled up or down almost without limit.

It also can be noted that electric potential is a linear function of resistivity. With this linear dependence of resistivity on potential, given a specified source current and associated potential, if all of the resistivities of the system are multiplied by the same scale factor, there is then a corresponding multiplication of potential by the same scale factor. This makes it possible to refer to a normalized set of apparent-resistivity and IP data for a specific modeled subsurface structure and then to scale the resistivities and correspondingly the measured voltages to match an actual physical condition, thus allowing a general solution to a specific geometry to be applicable to an infinite number of actual field cases.

The computation of theoretical IP and resistivity curves

The three-dimensional form of Ohm's law, $\vec{E} = \rho\vec{J}$ applies to the computation of theoretical IP and resistivity curves. Also, another law of physics requires that electric charge be conserved and that the net outflow of current from any volume must be zero. This law cannot be violated unless a current source is enclosed. The net outflow of electric current from a volume is measured by the divergence operator:

$$\nabla \cdot \vec{J} = \frac{\partial Jx}{\partial x} + \frac{\partial Jy}{\partial y} + \frac{\partial Jz}{\partial z} \tag{95}$$

where $\vec{J}$ is a vector with components J_x, J_y, and J_z, and ∇ is the gradient operator:

$$\vec{\nabla} = \frac{\partial}{\partial x}\vec{i} + \frac{\partial}{\partial y}\vec{j} + \frac{\partial}{\partial z}\vec{k} \tag{96}$$

Because an electric field is conservative for time-independent currents, its strength is derivable by a gradient operation from a scalar potential V as:

$$E = -\nabla V \tag{97}$$

Now $\nabla \cdot E = \nabla \cdot (\rho\vec{J})$, and assuming ρ to be uniform, $-\nabla \cdot \nabla V = \rho\nabla \cdot \vec{J}$ and when $\nabla \cdot \vec{J} = 0$, $\nabla^2 V = 0$. This is the well-known Laplace's equation of potential theory which governs the variation of electric potential V for a given distribution of resistive material.

Another boundary condition pertaining to potential fields was noted in eqs. 21 and 22 of Chapter 2 whereby current flow lines are refracted at a boundary between materials of contrasting resistivity. These equations can be rewritten as:

$$\frac{1}{\rho_1}\frac{\partial V_1}{\partial \mu_1} = \frac{1}{\rho_2}\frac{\partial V_2}{\partial \mu_2} \tag{98}$$

where μ is a direction cosine. The boundary conditions combine with $\nabla^2 V = 0$ to specify the complete and unique distribution of electric potential and current flow in and near a subsurface body.

There are several mathematical techniques that can be used to obtain theoretical solutions for the distribution of electric potentials from which master curves can be derived. These techniques can be called image solutions, series solutions, and integral solutions. All give the same numerical results but by different physical and mathematical approaches. The image solution method has been discussed in Chapter 2. Series solutions extend this method to an infinite number of images, and they are discussed in the reports by Ludwig and Henson (1967) available from Heinrichs Geoexploration Co. Integral solutions are derivable from the series solution method. They are mathematically attractive even though their physical meaning is not simple to grasp.

The derivations of the theoretical electric potential at a point on the earth's surface due to a point source of current have been given by Sunde (1949) and Koefoed (1968) among others. These formal analytic integral solutions are applicable to a layered earth with multiple horizons, and they are in the form:

$$V = \frac{\rho_1 I}{2}\frac{1}{r} + 2\int_0^\infty K(\lambda) \cdot J_0(\lambda r) \cdot \mathrm{d}\lambda \tag{99}$$

where ρ_1 is the surface resistivity, $K(\lambda)$ is a kernel function (Slichter, 1933), and $J_0(\lambda r)$ is a Bessel function, r is a distance from the current source, and λ

is a complex variable. Equation 99 can be solved numerically, allowing the otherwise laborious computations to be programmed on a digital computer.

Solutions presenting the surface potentials over a layered earth are usually shown in graphic form as families of apparent polarization and resistivity curves. The interpreter compares field data plotted on the same coordinate scale with the theoretical curves to find the best match. Knowing the value of the surface resistivity ρ_1, it is possible to determine the depth to a layer of contrasting resistivity and the resistivity of that layer. But these interpretations are not necessarily unique.

Master curves

In 1931, Roman introduced the graphic interpretation technique of curve matching using dimensionless scales as has been illustrated in Chapters 2 and 8. By proper superimposition of the plotted field data on a derived theoretical curve, the correct depth and contrast factor can be directly found. The method is simple, rapid, and accurate, but the interpreter must have a rather complete set of theoretical curves at hand. Also, the geometry of the subsurface layer must be assumed to some extent and slight inconsistencies in the data may be erroneously attributed to subsurface resistivity changes.

A large number of resistivity master curves have been published, but not many IP master curves are to be found in catalog form. The most complete index of IP curves has been computed by C.L. Elliot of Tucson, Arizona. An example of these curves is given in Chapter 8, Fig.8.10.

It is interesting and important to note that, although most potential field geophysical exploration methods are non-unique in solving for geometry and property contrast of a subsurface body, the solutions to horizontally layered resistivity problems are unique. This is an exciting challenge to the mathematician and interpreter alike. Caution is necessary in deciding the reason for any curve mismatch because there may actually be errors in the assumed boundary conditions related to lateral resistivity, inhomogeneities, lack of horizontality, and errors in field data.

The curve-matching technique gives an estimate of depth to interfaces and of physical property contrast. Usually, the field data curves are normalized to the value of the near-surface resistivity, which is presumed to be known from the closest electrode spacing. Most interpreters prefer the Schlumberger or Wenner arrays for making a vertical electric sounding (VES) for accurate resistivity estimations of depth. Electrode positions are usually expanded by 50% increments in order to provide a sufficient number of data points.

Depth estimates and physical property contrast interpretations can be made from IP data using the master curve method, but reconnaissance field survey data are often too widely spaced for accurate depth interpretations. Most routine IP surveys employ the dipole—dipole array with an incrementally increasing electrode spacing of $n = 1, 2, 3, 4, 5$, which does not provide very much data for a good depth estimate. Whereas resistivity VES depth estimates may be accurate to within 10%, the interpreter of reconnaissance IP data may have to be satisfied with inaccuracies of 25% or more. However, these depth inaccuracies may not be a hindrance in carrying out an exploration program because routine surveys can seldom be used to explore much deeper than the electrode interval itself and 25% of a 1,000-foot electrode spacing is within a reasonable estimation range. Furthermore, a reconnaissance IP survey is more often intended to discover a mineral deposit than to provide accurate depths to mineralization.

Because the IP property of mineralized materials is being contrasted with that of barren host rock, it is not too difficult to give a good estimate of the intrinsic polarization using the curve-matching method. Inaccuracies of up to 25% of the true polarization should be acceptable to the explorationist. Larger inaccuracies may be expected in weaker IP responses because of the presence of magnetotelluric noise or electromagnetic coupling.

To make a depth estimate by curve matching, a pseudosection of data is first examined for trends and patterns. Often the data are contoured with a logarithmic or geometric contour interval because of the large contrasts in electrical physical properties that exist in nature, seen especially in resistivity data. The interpreter may find it helpful to shade or color certain intervals between contours in order to bring out patterns and to facilitate rapid visual comparison of different sections of data.

To make a depth estimate for a horizontal layer, note a region in which pseudosection contours are horizontal and parallel over a length of several electrode intervals. Then plot the apparent resistivity as a function of electrode spacing on transparent graph paper at the same scale as the theoretical curves (Fig.2.23). Some averaging of data along the horizontal direction may be necessary. Draw a smooth curve through the data points and overlay the transparency onto the master curves. Move the transparency back and forth until it matches a comparable theoretical curve. Three-layer resistivity master curves are also available.

Direct interpretation

Instead of making a visual and manual comparison of apparent resistivity and IP field data with theoretically derived curves, it is possible to take the observed potentials and arrive at a direct solution of the subsurface resistivity structure. This method has been pointed out by Madden and his MIT students.

Direct interpretation of resistivity and IP data has not been commonly done because it has been tedious to digitize data and the computer operations have not been routine for a field geophysicist. However, with data now being processed in digital form and computers being actually used in the field, it appears likely that a direct interpretation system, interactive with the interpreter, will become a reality.

Numerical modeling

Although the earth is horizontally layered in many exploration situations, there are often lateral changes in resistivity and polarization that cannot be interpreted by the master curve comparison method. This has led to the investigation of mathematical modeling methods for different subsurface geometric shapes using various numerical modeling techniques. These methods have become more and more popular with the development of digital computers. Aiken and Hastings (1973) have a good review of this subject.

Two of the most versatile numerical modeling methods are the finite-element and finite-difference techniques. The basic concepts are that low-frequency current in conductive material is distributed so that power dissipation is minimized. For a closed volume without current sources or sinks (Coggon, 1971), the variation equation is:

$$\Delta\phi = \Delta\int \overline{J}\cdot\overline{E}\,dv = 0$$

and with a current source term $\vec{J}_s$ this becomes:

$$\Delta\int[\sigma(\nabla\phi)^2 - 2\overline{J}_s\cdot\nabla\phi]\,dv = 0$$

where $\overline{J}_s$ is current density, $\overline{E}$ is electric field, ϕ is potential, and σ is conductivity; dv is a volume element.

An expression must be found for the potentials such that the integral is minimized. This is done by dividing the volume into elements and applying the appropriate boundary conditions. The solution involves the manipulation of large matrices with many elements and nodes, so economics and computer capabilities must be considered in determining the spacing of the elements.

Finite-difference methods are based on the Taylor series expansion of a function of two variables. The converging series is truncated after an appropriate number of terms to obtain a solution to Laplace's equation. Boundary conditions provide constraints to the problem. The resulting system of linear equations can then be solved for the unknown potentials.

Finite elements have the advantage of irregular nodal spacing and nodal geometry so that irregular boundaries and heterogeneous models can be considered. Variably shaped elements allow a closer representation of model geometry and therefore fewer equations are needed than with the finite difference method. However, finite differences can handle three-dimensiona problems more simply.

GRADE ESTIMATION OF MINERALIZATION USING THE IP METHOD

In some situations, the resistivity and IP method can provide an estimate of the volume percent of mineralization of a subsurface body. However there are a number of uncertainties involved so that even under the best o

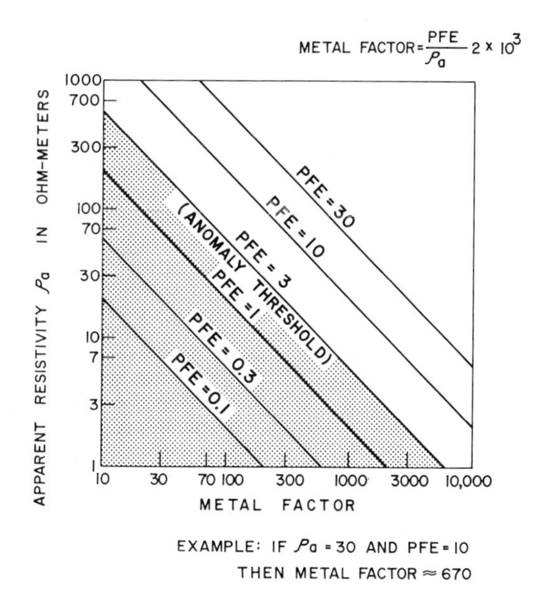

Fig. 9.1. Percent frequency effect (PFE) plotted as a function of metal factor and resistivity.

conditions the inaccuracy may be very large. The value of the grade of mineralization is obtained from the absolute intrinsic value of the electrical property of the body.

In the first place, some idea of the size, extent, or thickness of the subsurface body must be known because the product of the size and electrical property is obtained from survey data, but in order to separate the two variables, the value of one must be known.

Some interpreters have used chargeability as a guide to grade estimation, while others have used metal factor. Since these two parameters are quite different, any comparison of the methods cannot always be consistent. In fact, most grade estimates are based on the electrical characteristics of material of a known grade, pointing to the empirical nature of this kind of interpretation. Actually, grade estimations of this sort can be quite good, if the geological environment is relatively simple. Anaconda used the specific capacity parameter in estimating grade at Yerington, Nevada (Ware, 1973) and Bacon (1965) used metal factor in the copper country of Michigan, whereas Brant (1964) believes that chargeability worked best at Cajone, Peru. Interpreters of time-domain IP data have sometimes used the rule-of-thumb that one percent of sulfides will increase chargeability by 20 milliseconds, while frequency-domain interpreters have been known to estimate the sulfide grade by dividing PFE by three.

Wait (1959) and others have pointed out that particle size is also an important factor in polarization response. The presence of mineralized veins and the fabric of anisotropic rock can likewise influence the magnitude of polarization response, sometimes creating large errors in sulfide grade estimates made from IP data.

The metal factor must be used with considerable caution in interpretation, especially in low-resistivity areas. Where there is a high background of polarization, the significance of the metal factor as a polarization parameter is also questionable. These problems are illustrated in Fig.9.1.

QUALITY OF THE INTERPRETED DATA

As a general rule, an interpreter first familiarizes himself with the exploration history of the area, taking into account the physical properties of rocks, the exploration equipment, and survey method, and finally he reviews the data. If possible, the original field notes should be at hand so that the quality of the values can be determined. Good data are not difficult to verify if repeated field checks and calibration values are included. Questionable data do not contain such substantiation and must be suspect until proved otherwise.

It is well to keep in mind the possible sources of error in IP data, as mentioned in Chapter 1, and to make an estimate of the magnitudes of error. Unfortunately, there is a considerable amount of IP data of questionable quality that has been produced, most of which was obtained by outdated equipment. Interpretation is uncertain enough without using questionable data. In his report, the interpreter should make a statement of his evaluation of the data so that uncertainties will not be perpetuated.

REVIEW OF PSEUDOSECTION DATA

As a general rule, resistivity data are more closely related to the subsurface geology, including porosity, fracturing, structure, and rock type, than are IP data, and the IP data give a representation of the distribution of metallic mineralization in the subsurface. Thus, pseudosections for resistivity and for induced polarization must be reviewed separately and from different points of view. For example, the resistivity pseudosections will give the thickness of overburden or depth of oxidation, which can be of considerable value to the explorationist, while the IP pseudosections provide information about mineralization.

Most potentially mineralized areas have lateral variations in subsurface electrical properties, and these are displayed in a subtle fashion on the plotted IP pseudosections. The patterns of the pseudosection contours can be compared with modeled results to give an idea of intrinsic IP value, depth, and width. It is significant to note that due to limitations of resolution of the method the values for IP and resistivity effects are strongly attenuated

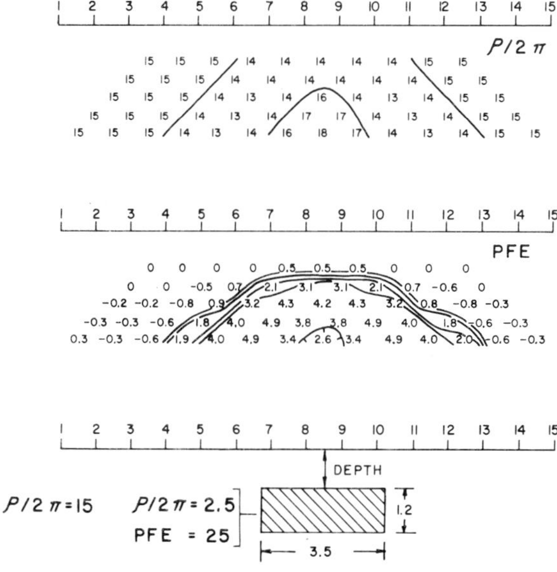

Fig.9.2. Resistivity and IP model pseudosections of a polarizable block at a depth of one electrode interval. Note the typical 'pant-leg' pattern of the weak resistivity low.

with depth. Figure 9.2 shows that a block of material with a true frequency effect of 25% buried at a depth of one electrode interval is only barely detectable from a surface survey.

THE INTERPRETATION REPORT

The interpreter is aware of the high cost of IP surveying and the even greater expense of exploration drilling, and he must budget an appropriate fraction of his time to the comprehensive discussion of the results. All aspects of the data must be reviewed and presented and recommendations made. Frequently, additional IP surveying should be done to contain the anomalous area. The reasons for drilling must be cited, together with proposed drill-hole locations, inclinations, and suggested minimum depths.

Since an interpreter's report is liable to be widely distributed and may become a valuable document, it deserves his best effort in preparation and presentation.

It is desirable that the interpreter be able to take advantage of the ongoing results of exploration information, particularly in the early stages of drilling a prospect. By incorporating these geologic data in the interpretation, he may be able to suggest modification of drill-hole locations or depths that could be of considerable benefit to the exploration program.

Chapter 10

THE COMPLEX-RESISTIVITY METHOD

Conventional IP data also contain essentially direct-current resistivity information, but if a maximum amount of electrical knowledge is desired, it is necessary to use additional alternating-current frequencies, looking at both the in- and out-of-phase components. This means that all of the complex (real and imaginary) resistivity characteristics measured over a spectrum of frequency values must be obtained. Such measurements, known as complex-resistivity determinations, require a comparison both in magnitude and phase of the output waveforms with the input waveforms, using both the frequency and time domains with an intervening transform. This will permit examination of the entire linear ground electrical system, as illustrated in Fig.1.3a.

The complex-resistivity (CR) method is therefore a variable-frequency, ground-contact method of studying the electrical properties of the earth by which both the in-phase and out-of-phase components of the earth's total resistivity are analyzed. The complex-resistivity method is a major step forward in electrical geophysical exploration inasmuch as it perceives the different parameters of the induced-polarization, resistivity, and related electrical and electromagnetic properties of the earth, separates their individual contributions, and presents a display for visual analysis and interpretation. It is not easy to separate these different parameters taken in a single broad range of measurements without the use of advanced electronic measuring devices and interfacing computational equipment. Although conventional IP survey may be used in the field for many years to come, the future trend in electrical surveying seems to be in the direction of complex-resistivity surveying.

HISTORY OF THE METHOD

When the approach, logic, and potential scope of the complex-resistivity (CR) method are appreciated, one may wonder why the technique has not been more widely used. The answer is that equipment development and theoretical understanding are still catching up with the technological advances. The CR method is new, and the first mention in the geophysical literature was by Van Voorhis et al. in 1973. Kennecott Exploration Services under Holmer started looking into the method in 1967. In 1973, Van Voorhis et al. reported on preliminary findings, the results of which appeared to be inconclusive. There were difficulties with equipment and technique, and not many rock samples were studied. Van Voorhis concluded that all rocks had a response similar to that shown in Fig.1.6, a feature that could be used for the elimination of electromagnetic coupling effects. This was, in fact, a breakthrough, because heretofore coupling was a serious IP surveying problem. But at that time there did not appear to be any variation in the rather flat spectral character of the rocks.

McPhar Geophysics developed two low-frequency multimode IP phase systems (Hallof, 1973, 1974), which used similar measurements of phase response to remove EM coupling. This technique was based on generalized (homogeneous-earth) theoretical coupling models. In 1972, Zonge had apparent success in measuring and isolating coupling and mineral response, and he thought there was a possibility that discrimination could be made between different rock types and between different metallic minerals.

One reason for optimism for success of the complex-resistivity system lies in its similarity to the revolutionary developments in reflection seismology which have resulted from equipment miniaturization, application of information theory, improved field methods, and computerization. A large amount of information is compiled whose electrical meanings are still being researched.

THE COMPLEX-RESISTIVITY SURVEY

The complex-resistivity survey can be looked at as a more refined form of the conventional IP survey in which an almost continuous frequency spectrum is measured instead of only a few basic output voltage magnitudes. The more recent CR systems developed by Zonge measure effects of discrete frequencies from 0.01 to 11.0 Hz, using odd harmonics (1, 3, 5, 7, 9, 11) of full square waves with frequencies at decade intervals (0.01, 0.1, 1.0, and 10.0 Hz). The use of the eleventh harmonic provides a frequency overlap with the next decade so that data can be checked for linearity.

An on-site computer commands the transmitter, samples the input waveform, and compares it with the sampled received waveform by employing a fast Fourier transform (FFT). Through the transform, the system truly operates in both the time and frequency domains, with phase angle being derived from the in-phase and out-of-phase components.

The relationship between conventional frequency-domain induced polarization and complex resistivity was shown on Fig.3.22. The data are plotted in the complex plane with the real axis on the abscissa and the imaginary axis on the ordinate, giving an almost continuous spectrum, with the lowest frequencies starting from the right portion of the spectral curve. This plotting method displays a maximum amount of interpretable information.

Without phase information, only absolute values of the voltages, that is, the lengths of the voltage vectors, can be known. The frequency effect is essentially an in-phase measurement, the absolute value of which is very nearly the value of the real (in-phase) component. Phase angle and chargeability are more an out-of-phase measurement. It can be asked if patterns in plotted measured values are interpretable as significant contributions resulting from the nature of the rock and its mineralogic components. There may indeed be empirical correlations, but exactly why plotted complex-resistivity data display these conditions is not yet well understood.

While chargeability and phase-angle patterns usually agree quite closely with one another, frequency effect may differ, especially where there are artificial conductors present, such as fences or pipe lines. Thus by measuring phase or chargeability and frequency effect, an idea can be obtained about the presence of cultural coupling in the data. This is illustrated in Fig.10.1, where there are three pipe lines and a lead-sheathed telephone cable crossing the survey area. Another way of identifying fences, pipe lines, and other artificial conductors is by their nonlinear response characteristics; that is, the

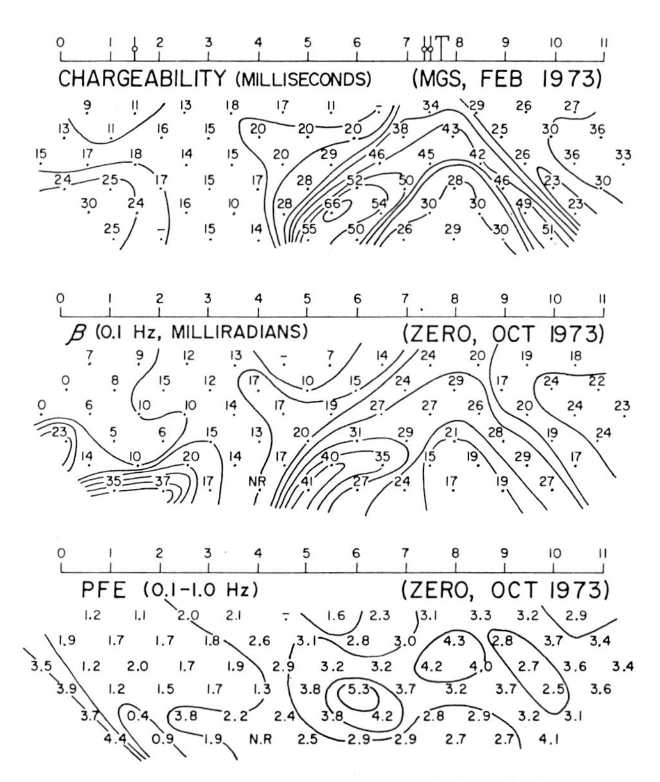

Fig.10.1. Chargeability, phase angle, and frequency effect pseudosections showing effects of cultural coupling.

```
  I
XMTR-I= 1.
 *G
GAIN= 120.
 *F
FREQ-E  10.0HZ
 *A
AVGS= 1024

1  0.9696  0.0294
3  0.9512  0.0249
5  0.9433  0.0231
7  0.9379  0.0223
9  0.9344  0.0201
;  0.9301  0.0190
 *F
FREQ-A  1.0HZ
 *A
AVGS= 128
 MAV:-14 #SP:
1  1.0137  0.0481
3  0.9842  0.0379
5  0.9733  0.0338
7  0.9667  0.0315
9  0.9621  0.0300
;  0.9585  0.0289
 *F
FREQ-I  0.1HZ
 *A
AVGS= 16
 MAV: 3 #SP:
1  1.0951  0.0654
3  1.0481  0.0592
5  1.0290  0.0545
7  1.0174  0.0520
9  1.0093  0.0487
;  1.0029  0.0462
 RHO-A      PHASE
  443.0     59.7
PFE
     8.6
 *F
FREQ-M  .01HZ
 *A
AVGS= 4
 MAV: 27 #SP:
1  1.1894  0.0441
3  1.1462  0.0622
5  1.1233  0.0655
7  1.1082  0.0667
9  1.0978  0.0665
;  1.0874  0.0669
 *
```

Fig.10.2. Raw complex-resistivity field data as printed out by the teletype.

HZ	REAL	IMAG
0.01	0.9993	0.0369
0.03	0.9627	0.0515
0.05	0.9439	0.0543
0.07	0.9309	0.0548
0.09	0.9227	0.0542
0.11	0.9136	0.0542
0.10	0.9180	0.0548
0.30	0.8782	0.0497
0.50	0.8626	0.0460
0.70	0.8525	0.0437
0.90	0.8469	0.0415
1.10	0.8413	0.0393
1.00	0.8433	0.0398
3.00	0.8188	0.0312
5.00	0.8098	0.0277
7.00	0.8044	0.0255
9.00	0.8006	0.0241
11.00	0.7977	0.0229
10.00	0.7987	0.0242
30.00	0.7836	0.0205
50.00	0.7773	0.0192
70.00	0.7729	0.0185
90.00	0.7700	0.0167
110.00	0.7668	0.0160

```
0.075 **** 0.045 **** 0.030 ****
**** 0.179 **** 0.442 **** 0.487
        RHO-A=    260.9 OHM-METERS
 0.1HZ PHASE=    59.7
0.1-1.0HZ PFE=    8.6

PLOT IN MAG/PHAZE?
```

ROCK SAMPLE MEASUREMENT

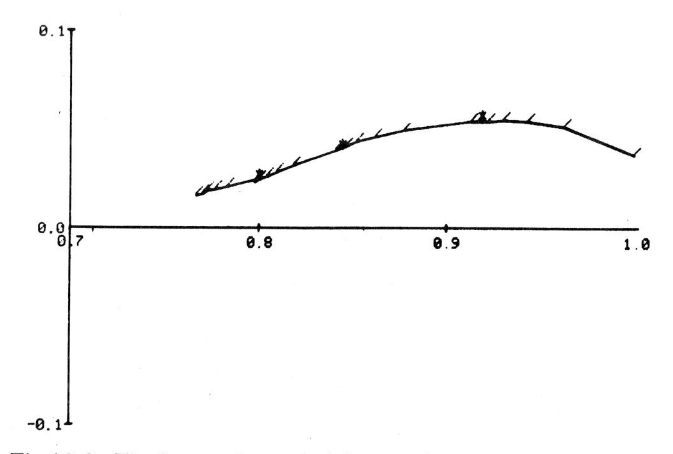

Fig.10.3. Final complex-resistivity results in the plotted and tabulated format. This is a measurement of a native copper sample showing a Type A response over three decades from 0.75 to 0.92 on the real axis.

HZ	MAG	PHASE	LOSS-TANGENT
0.01	1.0000	0.0369	27.1169
0.03	0.9640	0.0534	18.7037
0.05	0.9455	0.0574	17.3888
0.07	0.9325	0.0588	16.9823
0.09	0.9243	0.0587	17.0156
0.11	0.9152	0.0592	16.8586
0.10	0.9196	0.0596	16.7446
0.30	0.8796	0.0565	17.6731
0.50	0.8638	0.0533	18.7389
0.70	0.8536	0.0513	19.4891
0.90	0.8479	0.0489	20.4275
1.10	0.8422	0.0467	21.3820
1.00	0.8442	0.0472	21.1642
3.00	0.8194	0.0381	26.2410
5.00	0.8103	0.0342	29.2166
7.00	0.8048	0.0317	31.5609
9.00	0.8010	0.0301	33.2433
11.00	0.7981	0.0287	34.7821
10.00	0.7991	0.0303	32.9796
30.00	0.7839	0.0262	38.2008
50.00	0.7775	0.0247	40.5048
70.00	0.7731	0.0240	41.7075
90.00	0.7702	0.0217	46.0594
110.00	0.7669	0.0209	47.7825

FILTER OUTPUT?

ROCK SAMPLE MEASUREMENT

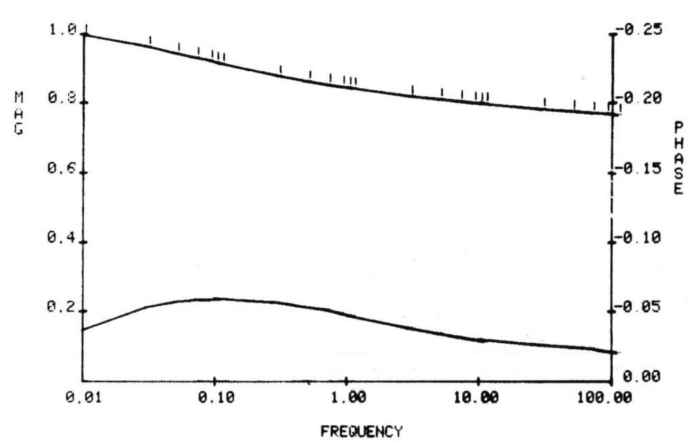

Fig.10.4. A magnitude and phase plot and table of results from the Fig.10.3 sample.

amount of polarization changes with the current density, a feature that can be seen in displayed complex-resistivity data.

The received waveforms are usually visually checked in the field, using a portable oscilloscope to monitor the voltage signal and noise. Then the computer is commanded to run through a spectral sequence and the portable field teletype prints out the results in the tabular form illustrated in Fig.10.2. The final results are presented in a plotted and tabulated display (Fig.10.3) or as magnitude and phase (Fig.10.4). These field data can be further enhanced in the laboratory to separate and interpret the spectrum and the electromagnetic coupling effects.

HOST-ROCK RESPONSE AND CR SPECTRA

It has been noticed that there is a predictable but as yet unexplained association between spectral signatures and different rock types. These signatures have been called Type A, B, and C responses and are shown in idealized form in Fig.10.5.

A Type A response shows a decreasing out-of-phase component with increasing frequency and is found where there is graphite, sulfide mineralization, or clay alteration of intrusive rock. A Type B response is characterized by a constant out-of-phase response for all frequencies, and it is often seen where there is moderate hydrothermal alteration and usually weaker sulfide mineralization than for a Type A response. A Type C response shows an increasing out-of-phase component with increasing frequency and is more often found with alluvium, limestone, or with chloritization of fresh volcanic rocks. These spectral trend characteristics are independent of the magnitude of the phase, chargeability, and frequency effect. But these trend effects may be resistivity dependent.

Separation of the electromagnetic coupling component in complex resistivity data

It appears to be possible to make a separation between actual polarization response of a rock and any electromagnetic coupling that may be present. The theory for analysis of spectral electromagnetic coupling has been presented by Wynn (1974). He notes that the EM coupling phenomenon is linear and that removal can be accomplished by an interactive digital computer operation using a cathode-ray oscilloscope display of the spectral curve (Fig.10.6).

The electromagnetic inductive coupling data, after separation from the polarization spectrum, can be analyzed in its own right to give an interpretation of subsurface conductivities and depths to interfaces. Also, there is an electromagnetic reflection coefficient parameter that can be identified which can be related to interfaces deeper than those normally detected in conventional IP resistivity surveys.

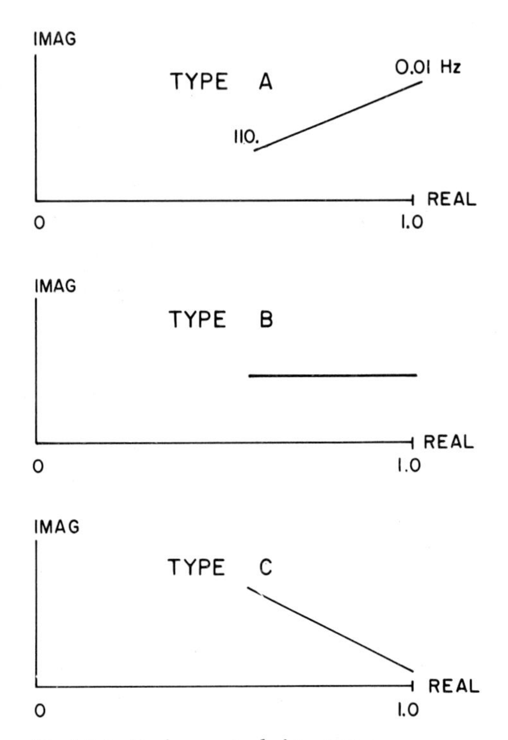

Fig.10.5. Basic spectral signatures.

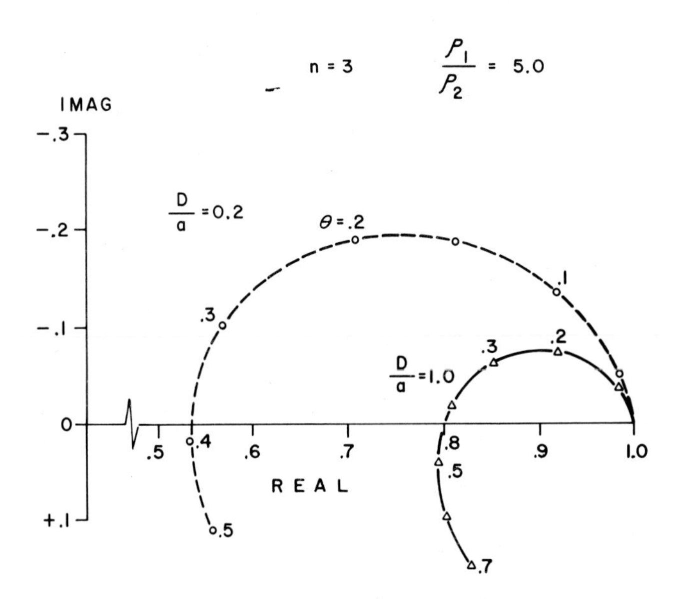

Fig.10.6. Theoretical EM coupling effects of a collinear dipole—dipole array over a layered earth. After Wynn (1974).

Identification of type of metal ion

Based on a large number of rock sample measurements and field survey CR spectral plots, Zonge (1972) believes that the presence of certain of the common sulfide minerals can be seen in the data. In particular, the presence of chalcopyrite apparently causes a small positive anomaly in the 10-Hz response, which can be seen on Fig.3.22 toward the left end of the spectrum. Pyrite seems to give a small positive anomaly at about 1.0 Hz. The rock used to generate this spectral curve is from Bisbee, Arizona, and it contains about 5—7% sulfides by volume.

SIGNAL DETECTION CAPABILITIES OF THE COMPLEX-RESISTIVITY METHOD

The CR method compares the received (output) and transmitted (input) signals in magnitude and phase by digitally sampling and transforming the waveforms from a time series to discrete frequencies, using the fast Fourier transform. This transform process is a powerful means of filtering noise of unwanted frequency from the data even though it is not an electronic filter in the conventional sense, and this process rejects a considerable fraction of the noise voltages that plague conventional IP surveys by a process of averaging and stacking data.

DISADVANTAGES OF THE COMPLEX-RESISTIVITY METHOD

At the present time, the principal objection to the CR method is that an electrical connection is necessary between the transmitter and receiver in order to make an accurate comparison of waveforms because telemetering or timing linkages have not yet proved successful to accomplish the operation and processing of data. This means that an extra connective wire must be taken into the field. Also, the electronic field and computer equipment is more sophisticated than that used with conventional IP surveys, and the CR method therefore requires trained geophysicists both in the field and in the analytical laboratory to carry out operations, adjust equipment, and process and analyze data.

Thus, CR surveying is more expensive than competing methods, but this disadvantage may be outweighed by the larger amount of high-quality information obtained. Future changes may tip the balance in favor of the CR method, but at the present time it is used more for detail surveys in areas where there is special interest because of unusual ground characteristics due to high-conductivity mineralization or noise problems.

Perhaps the most serious disadvantage of the CR method is that the sophistication of its new approach to understanding the electrical properties of the earth requires more study to appreciate than some of the exploration fraternity have been able to devote to it. The way to overcome this problem is continuing education and experience.

GLOSSARY OF TERMS USED IN INDUCED-POLARIZATION AND RESISTIVITY EXPLORATION

A

AC COUPLED INPUT. See COUPLING.

ACCURACY. The capability of an instrument to follow a true (measured) value. Inaccuracy is the departure from the true value due to all instrument errors, including repeatability, drift, temperature effects, and other causes. See SENSITIVITY.

ACTIVE SYSTEM. A system or circuit which must be supplied with external energy to sustain operation. Often pertains to electronic elements, such as amplifiers and filters. Compare PASSIVE SYSTEM.

ACTIVITY (electrochemical). The relative tendency of a substance to enter into a reaction. In petroleum self-potential well logging, a difference in solution activity due to different equivalent concentrations of dissolved salts accounts for a SELF-POTENTIAL.

ADMITTANCE. The reciprocal of impedance, or the complex ratio of current to voltage in a linear circuit. The practical unit is the MHO.

ADMITTIVITY. See CONDUCTIVITY.

ADSORPTION. A physical (electrostatic) chemical process in which a thin layer of molecules becomes fixed to the outer surface of a solid. See FIXED LAYER.

AELOTROPY. More commonly called ANISOTROPY.

ALTERNATOR. A rotating electromechanical device for supplying alternating current.

AMMETER. An electrical instrument used to measure the passage of current, in amperes.

ANALOG MODELING. A method of studying IP and resistivity effects of subsurface bodies by comparison with the response of models. Field surveys are simulated by measurements in an electrolytic tank using known conductive or polarizable shapes.

ANION. A negatively charged ion.

ANISOTROPY. Variation in the value of a physical property depending on the direction along which it is measured. In foliated rocks, the true resistivity parallel to foliation $\rho_{\parallel}$ is less than the true resistivity measured perpendicular to foliation $\rho_{\perp}$. Anisotropy of induced polarization in rocks is less than anisotropy of resistivity. See COEFFICIENT OF ANISOTROPY.

ANODE. An ELECTRODE (or portion thereof) at which oxidation occurs and electrons are produced. The positive terminal of an electrolysis cell or the negative terminal of a battery.

ANOMALY. A measured deviation from uniformity in a physical property, often of exploration interest. An economically interesting IP anomaly is usually positive and greater than BACKGROUND (or the NORMAL EFFECT). An economically interesting resistivity anomaly is generally less than background.

APPARENT. As applied to percent frequency effect, chargeability, or metal factor, the measured value for a property assuming the ground to be homogeneous and isotropic as distinct from true value. The subscript a is frequently used to indicate that a quantity is apparent; for example, $(PFE)_a$, M_a, $(MF)_a$.

APPARENT RESISTIVITY (ρ_a). The measured resistivity of the ground that is assumed to be homogeneous and isotropic in contrast with true resistivity in which these conditions are only locally real. Apparent resistivity is related to the ratio of measured voltage V to applied current I by the geometric constant K of the electrode array. Apparent resistivity ρ_a can thus be given as:

$$\rho_a = K\frac{V}{I}$$

and usually has units of ohm-meters. See RESISTIVITY.

APPARENT RESISTIVITY CURVE. The plotted relationship between apparent resistivity and electrode interval. Apparent resistivity curves can be normalized and compared with type curves of normalized apparent resistivity for the purpose of interpreting resistivity and depth of lower layers. See NORMALIZED APPARENT RESISTIVITY.

APPLIED POTENTIAL. A resistivity surveying method of mapping electric potentials in an area or volume surrounding the region where current is applied.

ARCHIE'S LAW. The empirical relationship of rock resistivity to its porosity. The ratio of rock resistivity ρ_r divided by pore-water resistivity ρ_w is equal to porosity of the rock ϕ raised to some power $-m$.

$$\frac{p_r}{p_w} = \phi^{-m}$$

where m is an empirical constant, called the CEMENTATION COEFFICIENT, ranging between 1.3 and 2.5. It is approximately 1.4 for slightly consolidated sandstones and 2.0 for limestones and most rocks of low porosity. See FORMATION FACTOR.

ARRAY. The particular pattern of arranging a series of electrical contacts (usually called ELECTRODES) on or in the ground. An array of electrodes is called symmetric if there is geometric symmetry. Synonymous with configuration as applied to electrode arrangements. Figures G.1 through G.7 give apparent resistivity and geometric factors in ohms-length, length being in the units of the surface survey.

AB RECTANGLE ARRAY. Identical to gradient array. Commonly used by the French school. See Fig.G.1.

AZIMUTHAL ARRAY. See AZIMUTHAL SURVEY.

DIPOLE—DIPOLE ARRAY. An array in which one dipole (a connected pair of electrodes) sends current into the ground and the other dipole serves as the potential measuring pair. The separation between pairs is usually only a few intervals greater than the spacing within each pair, so the electrode pairs are not ideal dipoles in a physical sense. The dipole pairs are usually collinear, but other orientations are possible. Resistivity and IP data from this array are sometimes displayed on a plotted pseudosection, as shown in Fig.G.2.

DOUBLE—DIPOLE ARRAY. See DIPOLE—DIPOLE ARRAY.

GRADIENT ARRAY. An array in which there are two distant, fixed current electrodes, A and B, and a pair of potential-measuring electrodes that are used to traverse a rectangular area between them. Identical to AB RECTANGLE ARRAY and similar to the SCHLUMBERGER ARRAY. See Fig.G.1.

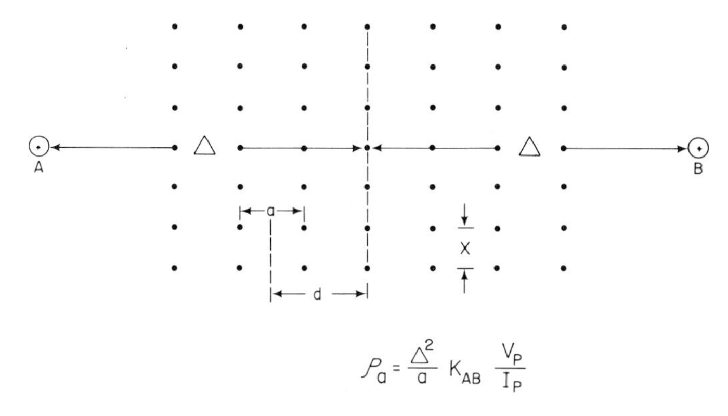

Fig.G.1. Plan view of the gradient array, or *AB* rectangle array. The rectangular area between distant, fixed current electrodes is traversed by a pair of potential measuring electrodes. Although the array factor K_{AB} is near unity, it is a variable.

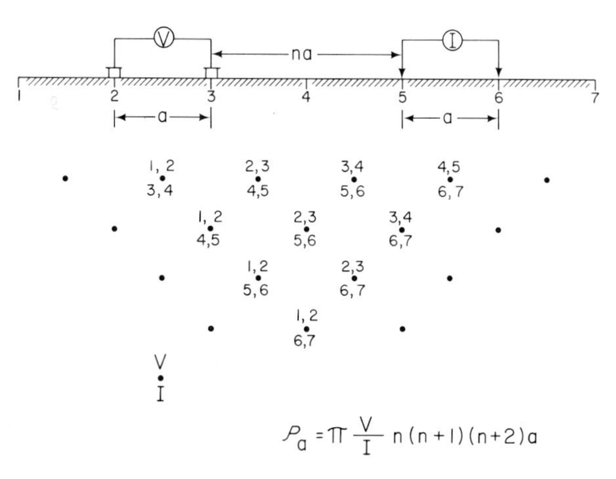

Fig.G.2. The dipole—dipole array and a pseudosection of plotted electrical survey data from expanding the array. Normalized data values are plotted at the intersections of 45-degree diagonals from the centers of the dipoles and trends are contoured.

POLE—DIPOLE ARRAY. An array in which the voltage-measuring pair of grounded potential electrodes is successively incrementally separated from one of the current electrodes (pole) while traversing a survey line. The second current electrode (the infinite electrode) is much farther from the first than is the dipolar pair of potential electrodes. Data can be plotted at the midpoint between the current and the two potential electrodes on a pseudosection. See Fig.G.3.

POLE—POLE ARRAY. An array in which two electrodes (poles), a current and a potential (with infinitely distant counterparts), are traversed or successively expanded on a survey line. Also called the two-array. Data are plotted

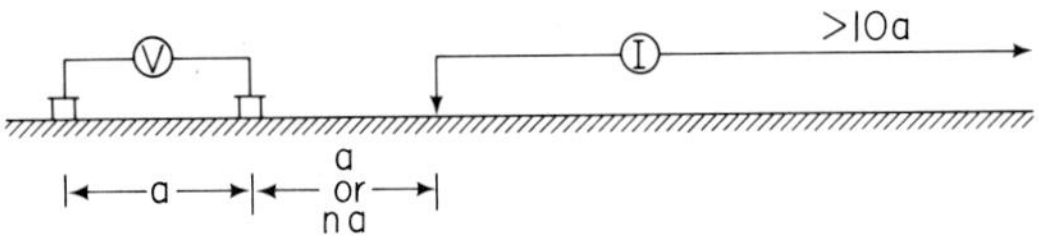

$$\rho_a = 2\pi \frac{V}{I} n(n+1)a$$

Fig.G.3. A pole—dipole array. This is called the three-array if the two potential-difference measuring electrodes and the nearest current electrode are equally spaced.

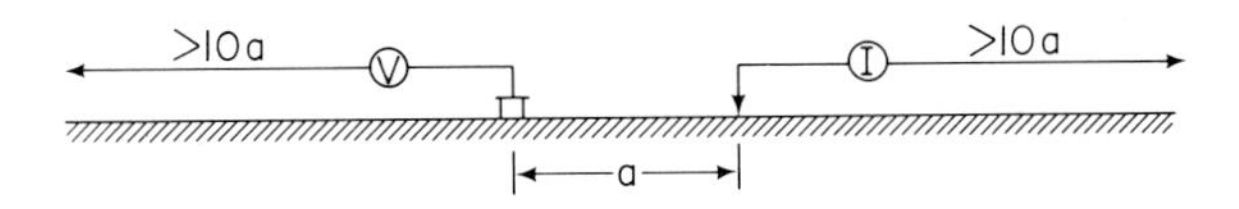

$$\rho_a = 2\pi \frac{V}{I} a$$

Fig.G.4. The pole—pole array.

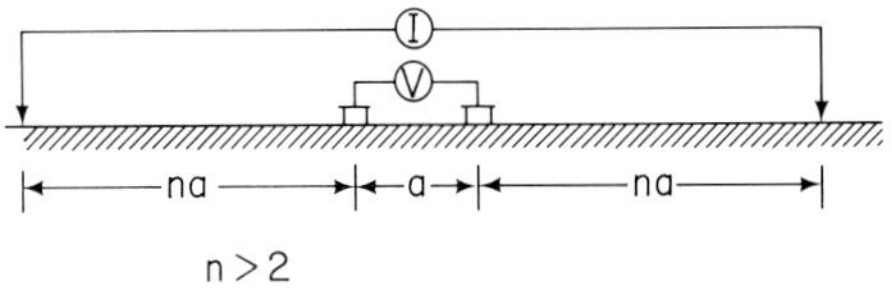

n>2

$$\rho_a = \pi \frac{V}{I} n(n+1)a$$

Fig.G.5. The Schlumberger array.

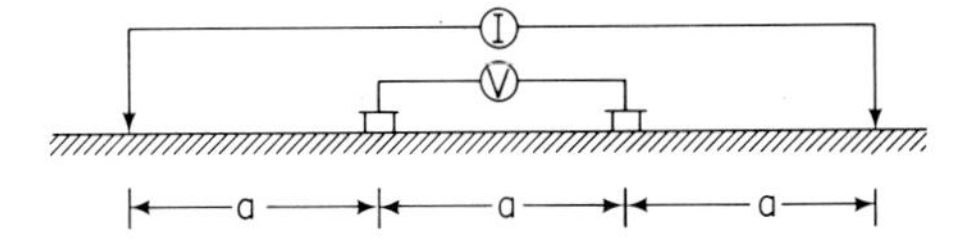

$$\rho_a = 2\pi \frac{V}{I} a$$

Fig.G.6. The Wenner array, consisting of four equally spaced in-line electrodes with the potential-difference measuring pair inside the outer current electrode pair.

either at the potential electrode or halfway between the two poles, as indicated on Fig.G.4.

RADIAL ARRAY. See AZIMUTHAL SURVEY.

SCHLUMBERGER ARRAY. An array in which the inner voltage-measuring pair of potential electrodes are much closer together (by a factor of about six) than the expanded outer current electrode pair. See Fig.G.5.

THREE-ARRAY. The array in which three equally spaced in-line electrodes and an infinite current electrode are used to traverse. The three in-line electrodes may be successively expanded. Measurements are plotted on a mapped position halfway between the near potential electrode and the closest current electrode or below this point if on a pseudosection. Similar to the POLE—DIPOLE ARRAY.

TWO-ARRAY. See POLE—POLE ARRAY.

WENNER ARRAY. An equal-spaced symmetric array, which can be expanded or traversed along a survey line. See Fig.G.6.

ARRAY FACTOR. See GEOMETRIC FACTOR.

ATTENUATION. A decrease in signal amplitude.

ATTENUATOR. A calibrated resistive circuit used on a voltage-measuring instrument so that voltages can be measured in a dynamic range of over several decade intervals of voltage. Sometimes called a multiplier or voltage divider.

AUTOCORRELATION. See CORRELATION.

AZIMUTHAL SURVEY. (1) A survey in which current electrodes laid out on the ground surface on specific azimuths from a drill hole are used to excite a potential at an electrode in the hole. The electrode in the hole can be successively raised to log potentials. On analyzing the potential data, the azimuthal direction toward better mineralization can be determined. (2) A survey method in which an area around a drill hole is traversed by a voltage-measuring potential electrode pair in radial directions (azimuths) away from a fixed current electrode, which may be down a deep drill hole. The second current electrode is at a distance infinitely greater than the interval between the potential pair, and it may be in a direction at right angles to the survey traverse. Also called RADIAL ARRAY. See Fig.G.7.

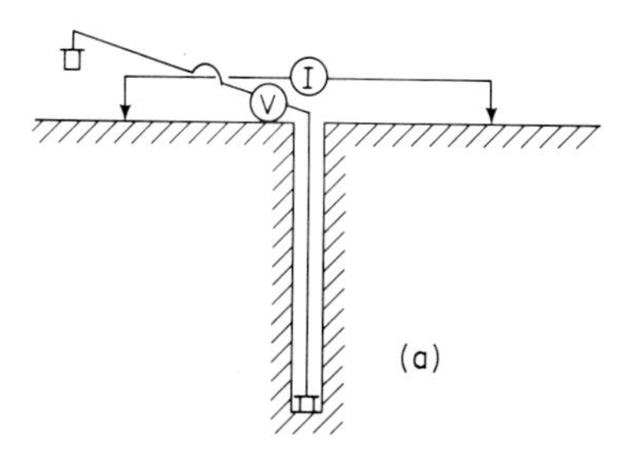

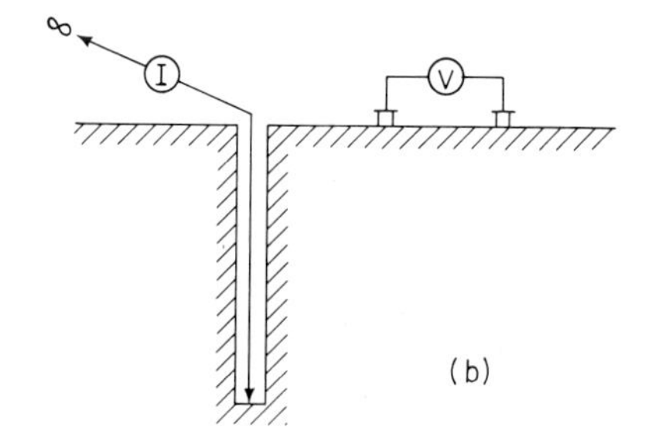

Fig.G.7. A cross-sectional illustration of variations of the azimuthal array, also called the radial array. (a) A deep drill hole with a potential electrode which can be successively raised during the survey. (b) A drill hole with a current electrode at the bottom and a potential pair used for surface surveying.

B

BACKGROUND POLARIZATION. The relatively weak IP response exhibited by many unmineralized rocks, particularly those containing abundant clay minerals or layered or fibrous minerals. Can also be due to weak, broad-scaled, pervasive mineralization which is not of economic interest. Also sometimes known as NORMAL EFFECT.

BALANCED INPUT. A symmetrical input circuit having equal impedance to ground from both input terminals.

BANDWIDTH. The range of frequencies over which a given device is designed to operate within specified limits.

BAY. A long-period transient magnetic disturbance, or magnetic storm, on an otherwise relatively undisturbed magnetic record. The onset of a bay is usually accompanied by a micropulsation burst, which accounts for some of the telluric noise observed in IP measurements.

BETA CURVE. A type curve used in interpreting IP data. Beta is proportional to the IP phase angle and is also a nondimensional loglog weighting factor developed from IP theory. For a simple single horizontal layer situation where only the lower material is polarizable, the resistivity contrast factor may be shown as a curve plotted against alpha (the ratio of array interval to depth) and beta. Beta is the ratio of observed apparent chargeability to the true chargeability of the lower medium. Also known as pulse curve or B FACTOR CURVE.

B FACTOR. A nondimensional bilogarithmic weighting function developed from IP theory and used to interpret the true IP property of a subsurface body. B factors for subsurface layers can be plotted as a curve. See BETA CURVE.

BIAS. In electronics, a voltage maintained at a point in a circuit so that the device will operate with optimal or other desired characteristics.

BIPOLE. A term referring to an electrode pair which must be considered to have a finite separation distance, in contrast to a physically true dipole which would have two electrodes relatively close together.

BIT. Abbreviation of binary digit. (1) A single character of a language employing exactly two distinct kinds of characters. (2) As used in connection with digital computing devices, a unit of storage capacity.

BLACK BOX. A unit or device whose basic function is specified but for which the method of operation is not specified.

BLOCK DIAGRAM. A diagram in which the important units of a system are drawn in the form of connected blocks. Often used to show the relationship of basic units of an electronic system.

BOUND LAYER. See FIXED LAYER.

BOUNDARY CONDITIONS. Known values of an otherwise sought-for solution at the boundaries of a domain. When time is a variable, the situation at $t = 0$ is an initial condition.

BOUNDARY VALUE PROBLEM. A differential equation together with boundary conditions.

BREADBOARD. An electrical circuit temporarily connected to test a new circuit or instrument before its design is finalized.

BRIDGE. One of several types of electrical networks containing a branch

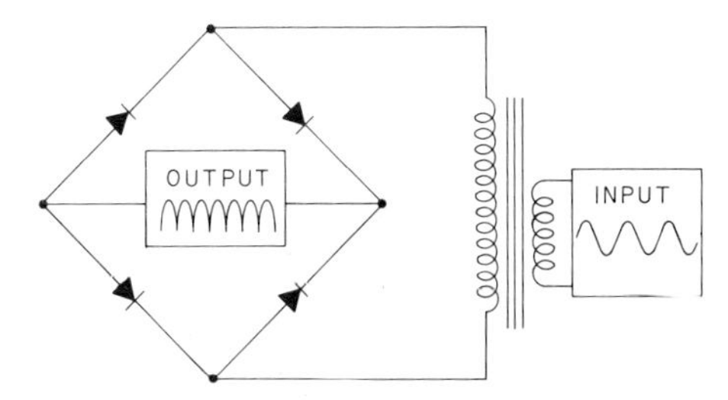

Fig.G.8. A full-wave rectifier circuit, consisting of four arms, each containing a rectifier, connected to an alternating current source through a transformer.

(the bridge) connecting two points of known potential for balancing. Sometimes used for measuring electrical impedance (impedance bridge).

BRIDGE RECTIFIER. A full-wave rectifier in which there are four arms, each containing a rectifier. See Fig.G.8.

BUGS. Errors or malfunctions in equipment, computer program, etc.

C

CALIBRATION. A check of equipment made by making a measurement on a known standard. In induced polarization, voltmeter receivers must occasionally be checked, or calibrated, with an IP transmitter using a non-polarizing circuit or a circuit with known polarization characteristics that are similar to those of the earth. This process often gives a small receiver voltage correction factor, called a calibration factor, that must be applied to field data. IP transmitters sometimes also need to be aligned.

CALIBRATION FACTOR. See CALIBRATION.

CALIBRATION RESISTOR. A pure resistance of known value used to calibrate the zero frequency effect level of a frequency domain transmitter and receiver.

CAPACITANCE (C). The ratio of charge Q on a capacitor or condenser to the corresponding change in potential V. Also the property of collecting an electric charge. $C = Q/V$. The units of capacitance are farads.

CAPACITIVE REACTANCE (X_c). Electrical impedance due to capacitance C, expressed in ohms, $X_c = 1/2\ \pi\ f\ C$, where f is frequency. See REACTANCE.

CAPACITIVITY (κ_0). The three-dimensional capacitance of a material, measured in farads per meter. Free space has a capacitivity of $8.854 \cdot 10^{-12}$ F/m. Also called PERMITTIVITY or DIELECTRIC CONSTANT. The IP effect can be viewed in terms of capacitivity.

CATHODE. The electrode at which reduction occurs and electrons are taken up. The negative terminal of an electrolysis cell or the positive terminal of a battery. Also, the source electrode for electrons in a vacuum tube.

CATHODE-RAY OSCILLOSCOPE (CRO). A waveform viewing device, measuring voltage as a function of time. Useful for testing and research.

CATION. A positively charged ion.

CATIONIC MEMBRANE. A membrane which permits the passage of cations but not of anions. Shale acts as such a membrane, allowing sodium ions to pass but not chloride ions. This phenomenon is important in the generation of self-potentials (SP) as observed in petroleum well logs.

CEMENTATION COEFFICIENT. See ARCHIE'S LAW.

CGG. Compagnie Générale de Géophysique.

cgs. The centimeter-gram-second system of units.

CHARGEABILITY (M). The basic unit of induced polarization in the time domain.
(1) As often measured in the field, the integrated area under an IP decay curve between times t_1 and t_2, normalized by the primary voltage V_p. Units are millivolt-seconds-per-volt, or milliseconds, given as:

$$M = \frac{1}{V_p} \int_{t_1}^{t_2} V_t \mathrm{d}t$$

and shown on Fig.G.9. For purposes of standardization of measurement, the Newmont group (Wait, 1959) suggest that the number of seconds of on time and off time be indicated by subscripts; that is, ${}_{33}M_1$, or $M_{(331)}$, means current on three seconds and decay measured for the first second of a three-second off time. Field measurements of chargeability are usually calibrated to the $M_{(331)}$ standard, which is closely related in magnitude (× 1,000) to the theoretical value of M given in definitions (2) and (3) below.
(2) The ratio of initial decay voltage (or secondary voltage) V_s to primary voltage V_p:

$$M = \frac{V_s}{V_p}$$

as shown on Fig.G.10.
(3) According to steady-state IP theory, the dimensionless IP parameter M of a material in which there is an induced-current dipole moment per unit volume $\vec{P}$ energized by a current density $\vec{J}$:

$$\vec{P} = -M\vec{J}$$

(4) In differential terms, the ratio ΔV or $\Delta \rho$ measured on a decay curve and voltage V or resistivity ρ observed with time:

$$M = \frac{\Delta \rho_t}{\rho}$$

(see also G.11)

CHEMISORPTION. Adsorption due to chemical rather than to simply electrostatic causes.

CHOKE. An inductor designed to present a relatively high impedance to alternating current in an electrical circuit.

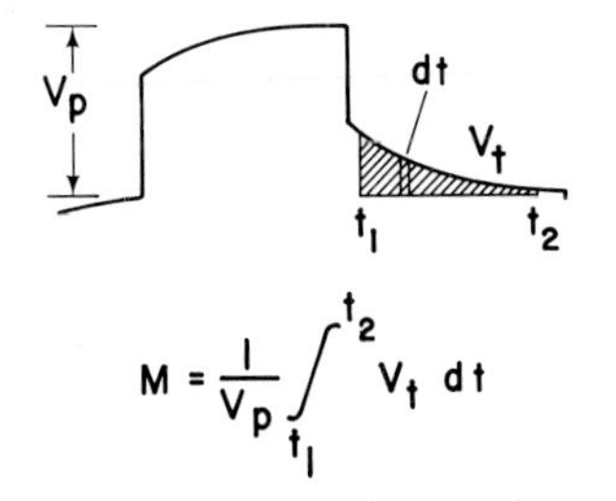

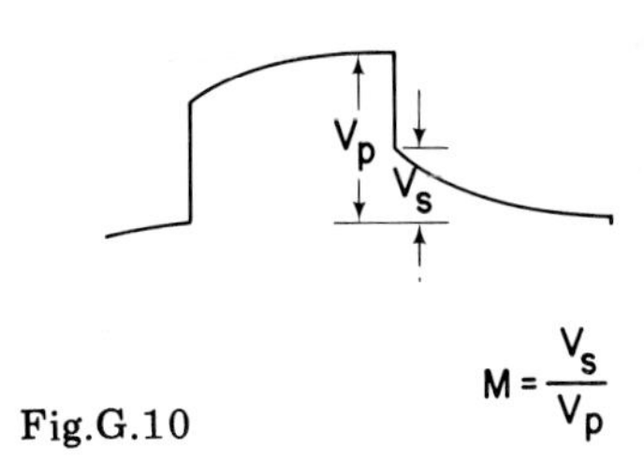

Fig.G.10

Fig.G.9. Chargeability measured as the area under the IP decay curve.

Fig.G.10. A charge-decay curve seen as a voltage due to an impressed pulse of current. An instantaneous measure of chargeability. $M = V_s/V_p$.

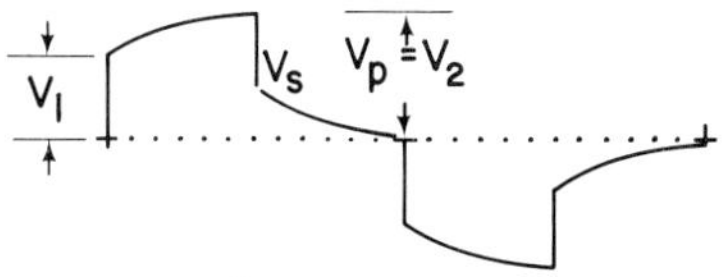

V_1 V_s $V_p \pm V_2$

$$M = \frac{V_s}{V_p} = \frac{\Delta V_t}{V_p} = \frac{V_2 - V_1}{V_2}$$

$$M = \frac{FE}{1+FE}$$

FOR SMALL FE, $M \approx FE$

FOR LARGE FE, $M \rightarrow 1$

Fig.G.11. The differential measurement of chargeability M in which a voltage increment ΔV on the decay curve is divided by the steady-stage voltage V. Voltage is proportional to resistivity so that $M = \Delta V/V = \Delta\rho/\rho$.

CHOPPER. An electrical switching device, sometimes including an oscillator, used, for example, to interrupt a dc or low-frequency ac voltage so that it can be filtered or measured by an ac voltmeter.

CIRCUIT BREAKER. A current-sensitive switch which opens (like a fuse) at high current levels to protect other circuit elements from the overload.

COEFFICIENT OF ANISOTROPY (λ). The square root of the ratio of the true transverse resistivity ρ_t to true longitudinal resistivity ρ_L in an anisotropic material:

$$\lambda = \sqrt{\frac{\rho_t}{\rho_L}}$$

The coefficient λ is always greater than one and is generally less than two.

COHERENCE. (1) A property of two in-phase wave trains. (2) A measure of the similarity of two time functions or portions of functions. If the functions have power spectra ρ_{ii} and ρ_{jj} and cross power spectra ρ_{ij}, their coherence is $\rho_{ij}/\sqrt{\rho_{ii}\rho_{jj}}$. Coherence is a concept in the frequency domain analogous to correlation in the time domain.

COHERENT DETECTION OR FILTERING. A method of noise suppression that relies on the comparison of two time-related signals. Synchronous detection and common mode rejection methods are similar in principle. 'Coherent' refers to the in-phase measurement.

COLE—COLE PLOT. A diagram on which is plotted out-of-phase response as a function of in-phase response at successive frequencies.

COMMON MODE REJECTION. A measure of how well a differential amplifier ignores a signal that appears simultaneously at both input terminals.

COMMUTATE. To reverse mechanically the direction of a unidirectional electric current by changing connections, or to reverse every other cycle of an alternating current so as to form a unidirectional current.

COMPARATIVE INTERPRETATION. The comparison of electrical survey data with theoretical curves obtained over bodies of assumed electrical property contrast and geometry or geologic structures (Van Nostrand and Cook, 1966).

COMPARATOR. A circuit which compares two signals and indicates the result of the comparison. May be an integrated circuit (IC).

COMPLEMENT OF CHARGEABILITY (L). An IP time-domain measurement of the area over the decay curve, integrating voltage over the interval between 0.45 and 1.75 seconds. Can be measured on a Newmont-type IP receiver. Used with chargeability M in interpretation of data.

COMPLEX RESISTANCE. The impedance of a circuit. The vector sum of real (in-phase) and imaginary (out-of-phase) components.

COMPLEX RESISTIVITY. The real and imaginary resistivities of a three-dimensional material.

COMPOSITE DECAY CURVE. A time-domain voltage decay curve containing more than a single decay component, often with different time constants or even a combination of positive and negative decay curves.

CONDUCTANCE. With direct current, the reciprocal of resistance. With alternating current, the resistance divided by the square of the magnitude of the impedance, or the real part of the admittance. The practical unit is the MHO.

CONDUCTION ANGLE. The number of degrees in a half-cycle ac wave during which a silicon controlled rectifier (SCR) is turned on. If ϕ is the phase control angle, the conduction angle is $180° - \phi$.

CONDUCTIVITY (σ). A measure of the ease with which current can be driven through a material by an electric field, properly measured in mhos per meter. In an isotropic substance it is the reciprocal of resistivity. Also sometimes called SPECIFIC CONDUCTANCE.

CONDUCTIVITY SPECTRUM. See RESISTIVITY SPECTRUM.

CONDUCTOR. A body within which charges can flow freely.
(1) ELECTRONIC CONDUCTOR. A substance that conducts electricity primarily by means of electrons. A substance whose electron mobility is high. Most electron-conducting metallic-luster sulfide minerals are really semiconductors.
(2) IONIC CONDUCTOR. A substance that conducts electricity primarily by means of ions. A substance whose ion mobility is high. Electrolytes are ionic conductors.

CONFIGURATION. See ARRAY.

CONSTANT CURRENT. Refers to requirements of the inducing current source that is used to excite the IP effect.

CONTACT RESISTANCE. The resistance observed (1) between a grounded electrode and the ground; (2) between an electrode and a rock specimen; or (3) between electrical contacts.

CONTROLLED RECTIFIER. An electronic circuit element consisting of a controlled diode or solid-state switch. The controlled diode is usually turned on by a small voltage from an external circuit and turned off when the voltage is reversed. Used to switch large currents or to control large currents in IP transmitters. When the semiconductor is silicon, these controlled rectifiers are called silicon controlled rectifiers (SCR).

CONVOLUTION. (1) The change in wave shape as a result of passing a signal through a linear filter. (2) The act of linear filtering. Convolution in the time domain is multiplication in the frequency domain. See FILTERING.

CORRELATION. A measure of the similarity of two waveforms or the degree of linear relationship between them used in the time domain. Analogous to coherence in the frequency domain. Also called AUTOCORRELATION and CROSS CORRELATION.

COULOMB'S LAW. The inverse-square attractive and repulsive law dealing with the force between charges. $F = (Q_1 Q_2)/r^2$ where F is force, Q is charge, and r is the distance between charges.

COUPLING. Refers to the type of mutual electrical relationship between two closely related circuits. An ac coupling on the input of a voltmeter-receiver would exclude dc voltages by employing a series capacitive element back of one of the input terminals. A dc or direct coupling on the input of a voltmeter-receiver may exclude higher frequency ac signals by using a capacitive element across the inputs or may allow all components of a signal to pass.

CAPACITIVE COUPLING. Refers to mutual capacitive impedance, for example, between the grounded circuits used in IP surveying. Under these conditions, the coupling is due to capacitive or dielectric effects between wires or between a wire and the ground.

ELECTROMAGNETIC (EM) COUPLING. An electromagnetic phenomenon involving the higher frequencies or short-time, lower resistivities and greater distances between electrodes that gives rise to electromagnetic effects that can be mistaken for IP effects. Also called INDUCTIVE COUPLING or mutual inductive impedance.

INDUCTIVE COUPLING. The mutual inductive impedance that exists in IP field surveys between grounded transmitter and receiver circuits, especially at higher frequencies, greater distances, or lower earth resistivity. An undesirable condition that gives rise to false IP anomalies. Also called ELECTROMAGNETIC (EM) COUPLING.

RESISTIVE COUPLING. In IP surveying, this type of coupling is due to current leakage between wires or between a wire and ground. Also, resistive coupling is the dc ohmic effect of the ground itself between two grounded circuits used to measure apparent resistivity.

CRITICAL DAMPING. The amount of damping beyond which oscillatory motion ceases.

CROSS CORRELATION. See CORRELATION.

CURRENT DENSITY (J). The amount of current per unit area in a material; determined by the velocity and density of charge carriers. If too large currents are used for IP work, as in some laboratory measurements, resistivity of earth materials becomes non-ohmic or IP effects become nonlinear. Current density J is a vector quantity measured in amperes per square meter.

CURRENT ELECTRODE. A prepared, grounded contact of the IP transmitter circuit. Usually an attempt is made to give this contact a low electrical resistance in order to put the greatest possible current into the ground.

CURVE MATCHING. See COMPARATIVE INTERPRETATION.

CUTOFF. The frequency at which a filtered response is down by a predetermined amount, such as 3 dB. The cutoff points designate the filter: an 0.3—3.0 filter has a low-frequency cutoff at 0.3 Hz and a high-frequency cutoff at 3.0 Hz.

D

DAMPING. A means of slowing down or opposing oscillations through dissipation of energy within the system.

METER DAMPING. A method of damping the needle fluctuations of a voltmeter dial by inserting a resistive and perhaps capacitive element across the instrument, giving a time constant to the settling of the meter needle.

DAR ZARROUK. The name given by Maillet (1947) to resistivity parameters or curves that deal with layered anisotropic materials.

DC COUPLING. See COUPLING.

DC PULSE METHOD. See PULSE IP METHOD.

DEBUG. To examine, search for, and remedy malfunctions or errors in instruments or computer programs.

DECADE. As used in electricity, a multiple of a factor of 10 more or less than a given value.

DECAY CONSTANT (τ). The time for an exponentially changing voltage to vary by 1/e or to change 63% from its initial value. Also called TIME CONSTANT.

DECAY CURVE. A plotted curve described by the decay voltage V_t against time. An IP voltage decay curve may be characteristic of an earth material. In theory it can be mathematically transformed to a resistivity spectrum.

DECIBEL (dB). A unit used in expressing power and intensity ratios: 20 $\log_{10}$ of the voltage ratio or 10 $\log_{10}$ of the power ratio. A voltage ratio of 2, which represents a power ratio of 4, is equivalent to 6 dB. Resistances of the circuits being compared must be the same when compared by the dB ratio.

DEL (∇). The gradient operator, a vector, in rectangular coordinates:

$$\vec{\nabla} = \vec{i}\frac{\partial}{\partial x} + \vec{j}\frac{\partial}{\partial y} + \vec{k}\frac{\partial}{\partial z}$$

where $\vec{i}, \vec{j}$, and $\vec{k}$ are unit vectors. ∇U is the gradient of the potential field U. ∇ is also called NABLA.

DELAY TIME. In IP work, the time interval between the off instant of the charging current and the on instant of the measuring voltmeter oscillograph. Delay time of sometimes up to 450 or even 1,000 milliseconds is necessary to allow dissipation of transient voltages which are not directly related to the polarization decay voltage V_t.

DEPTH OF EXPLORATION. The depth below surface to which an exploration system can effectively explore. Depends on array, spacing, property contrast, body geometry, and signal-to-noise ratio. Also see SKIN DEPTH.

DEPTH PROBE. A technique of exploring successively downward into the earth by employing an orderly horizontal expansion of the interelectrode interval of an array. Data from such a depth probe survey (also called sounding or using an expander) can be interpreted to give a depth to a contrasting resistivity or anomalous IP material if horizontal layering exists. Also called depth profile or VERTICAL ELECTRIC SOUNDING (VES).

DETECTION LOGGING. See EXPLORATION LOGGING.

DIELECTRIC CONSTANT (K). A measure of the capacity of a material to store charge when an electrical field is applied. The capacitivity of a material. Also called PERMITTIVITY or dielectric coefficient.

DIELECTRIC LOSS. The energy loss per cycle in a dielectric material due to conduction and slow polarization currents or other dissipative effects.

DIELECTRIC POLARIZATION. The response of a dielectric material to the presence of an electric field, producing an induced dipole moment per unit volume. In an insulating dielectric material, no net electric charge need be transferred by the exciting field. By some definitions, induced polarization is a lossy-type dielectric polarization with a longer time constant.

DIFFERENTIAL. A term applied to electronic equipment, such as amplifiers and voltmeters, to describe a type of device that measures the difference in characteristics between signals. A differential input on a voltmeter is desirable to reject noise originating from the ground. See COMMON MODE REJECTION.

DIFFUSE LAYER. The outer, more mobile ions of an electrolyte—solid interface which together with the fixed layer constitutes a double layer. Also called diffuse zone, diffuse double layer, or outer Helmholtz double layer. See DOUBLE LAYER.

DIFFUSION. The motion of ions or molecules in a solution resulting from the presence of a concentration gradient.

DIFFUSION IMPEDANCE. The impedance in a solution produced by a concentration gradient, for instance, one next to an electrode. See WARBURG IMPEDANCE.

DIGITIZE. To sample a continuous function at discrete time intervals, quantize the measurements, and record the values as a sequence of numbers.

DIMENSIONAL ANALYSIS. The study of the equating units in a physical relationship in which the dimensions as well as the number values must balance.

DIP. Differential induced polarization, a two-electrode electrical prospecting method being promoted by Alex Lebodowsky. The exact electrical parameters being measured by this method are uncertain.

DIPOLE. A pair of unlike poles, such as electrical charges or poles of a magnet, that ideally are infinitesimally close together. In resistivity and IP surveying, a pair of nearby current electrodes approximates a dipole field from a distance and the use of a voltage-detecting electrode pair is a dipole-like probe.

DIPOLE MOMENT PER UNIT VOLUME (P). A measure of the intensity of polarization of a material. In IP, the units are ampere-meters per cubic meter.

DIRECT INTERPRETATION. Direct mathematical solution without use of precomputed curves or models. Used in the solution of potential field problems as encountered in resistivity and IP surveying. Directly treats data to give subsurface conditions. The English school (Tagg, 1964; Griffiths and King, 1965) use this term in CURVE MATCHING.

DIRECTIONAL SURVEY. An IP or resistivity survey method starting from a position, such as a drill hole, to find the trend direction of an anomalous subsurface body.

DIRTY SAND. A composite electrolyte system containing clay and sand, which can be responsible for membrane polarization effects. Clay particles in the sand act as selective 'ion sieves'. The resistivity is low because of the surface conduction along the clay minerals.

DISPERSION. (1) The separation of electromagnetic radiation into some variable of the radiation, such as energy or frequency. (2) A statistical term for the amount of deviation of a value from the norm.

DISPLACEMENT CURRENT. A concept accounting for the nonsteady-state flow of current through a nonconductor, such as the dielectric in a capacitor.

DISSEMINATED SULFIDE MINERALIZATION. Sulfide minerals scattered as specks and veinlets through the rock, constituting not over 20% of the total volume. Compare MASSIVE SULFIDES.

DISTRIBUTED. Referring to electric circuits, the 'smearing out' of circuit elements into a network, such as a ladder circuit, to give an exponential or logarithmic response to simulate, for instance, the Warburg impedance. Compare LUMPED CIRCUIT.

DISTRIBUTION FUNCTION. A statistical relationship used to describe the characteristics of a composite function, such as a decay curve, containing more than one time constant. The distribution function $G(u)$ is the density of time constants in the interval u to $u + du$ where u is the reciprocal of the time constant; $u = 1/\tau$.

DOMAIN. A set of parameters, one of which is an independent variable, such as frequency or time.

DOUBLE LAYER. The layer of molecular ions and charge dipoles at a solid-solution interface. It is electrically analogous to a capacitor in that there is charge separation between the solid (electrode) and the charge center of the oriented ions or dipoles. Next to an electrode there may be an adsorbed fixed layer of ions that is also called the inner Helmholtz double layer. A diffuse layer (outer Helmholtz double layer) in the electrolyte contains an excess of ions, which are usually of the same charge as the electrode but of opposite charge to that of the fixed layer. The thickness of the double layer is less than 100 Å. Double-layer effects are not significantly important at the IP frequencies normally used in the field.

DOUBLE-LAYER CAPACITANCE. Capacitance due to the presence of the double layer. The frequencies at which double-layer effects become important are higher (above 100 Hz) than those normally used in IP prospecting.

DOWN-THE-HOLE IP METHOD. Any survey method that explores the region near a drill hole using only a single potential or current electrode in the drill hole and other electrodes on the ground surface. Compare IN-HOLE IP METHODS.

DUMMY LOAD. A ground-matching resistance frequently used with pulsed square-wave transmitters to balance the output load from the transmitter during the power-off portion of the duty cycle.

DUTY CYCLE. The percentage of time in which current is delivered during a complete waveform cycle of an IP transmitter.

E

EAEG. European Association of Exploration Geophysicists.

ELECTRICAL DOUBLE LAYER. See DOUBLE LAYER.

ELECTRICAL PROFILING. See PROFILING.

ELECTRICAL SOUNDING. See SOUNDING.

ELECTRIC FIELD (E). A spatial vector quantity measurable as a potential gradient which may be due to a charged body or to a varying magnetic field. Unit is volt per meter.

ELECTRIC LOGGING. Drill-hole IP, resistivity, or SP measurements that are made in a drill hole. Results are shown on an electric log giving the electrical parameter(s) as a function of in-hole depth of the electrode array.

ELECTRIC SUSCEPTIBILITY (χ). The ratio of intensity of electric polarization P to electric field E. A measure of the polarization property of a dielectric. Also called dielectric susceptibility. The primary term of the METAL FACTOR.

ELECTROCHEMISTRY. The branch of chemistry that deals with the chemical changes produced by electricity and the production of electricity by chemical changes.

ELECTRODE. A piece of metallic material used as an electrical contact with a nonmetal. Can refer to a grounding contact used for field surveying, to the metallic minerals in a rock, or to electrical contacts in laboratory equipment.

ELECTRODE ARRAY. See ARRAY.

ELECTRODE EQUILIBRIUM POTENTIAL. The reversible equilibrium potential (no energy loss) across the interface between an electrode and an electrolyte, when no current is passed through the interface. It must be measured as the voltage difference between the electrode and a reference electrode. It is due primarily to the free energy of the electron transfer process.

ELECTRODE IMPEDANCE. (1) In electrochemistry, the total impedance across the interface between an electrode and an electrolyte; includes solution resistance, capacitances in the fixed and diffuse layers, and Warburg impedance. Interpretation of this impedance requires the assumption of a particular model electrical equivalent circuit. (2) In electrical circuit theory, the self-impedance of a single electrode or the mutual impedance between electrodes.

ELECTRODE POLARIZATION. (1) In electrochemistry, the concept that an electrode is polarized if its potential deviates from the reversible or equilibrium value. Polarization can be induced by passage of current through an interface or by a change in ion concentration at an electrode surface. The amount of polarization in excess of equilibrium is the overvoltage of the electrode. Also called overvoltage or induced polarization. (2) In IP field surveying, a term used to describe a diode-like behavior of electrode resistances of an electrode pair.

ELECTRODE POTENTIAL. See ELECTRODE EQUILIBRIUM POTENTIAL.

ELECTRODE RESISTANCE. The electrical resistance between an electrode and the adjacent material with which it is in contact. Sometimes called CONTACT RESISTANCE, SELF-RESISTANCE, or MUTUAL RESISTANCE, depending on the conditions.

ELECTRODIALYSIS. The phenomenon of migration of charge through a membrane in an electric field.

ELECTROKINESIS. A general term for dynamic electrical phenomena. Descriptive of electric potentials observed as a result of fluid movement. See STREAMING POTENTIAL.

ELECTROKINETIC POTENTIAL. See STREAMING POTENTIAL.

ELECTROKINETICS. The branch of physics that deals with electricity in motion.

ELECTROLYTE. A material in which the flow of electric current is accompanied by the movement of matter in the form of ions. Also any substance that dissociates into ions.

ELECTROLYTIC TANK. A container holding a conductive solution in which electrical model experiments can be carried out. See ANALOG MODELING.

ELECTROMAGNETIC (EM) COUPLING. See COUPLING.

ELECTRONIC. Relating to devices, circuits, or systems in which conduction is primarily by electrons moving through a vacuum, gas, or semiconductor.

ELECTRONIC CONDUCTOR. A material, such as a metal, that conducts electricity by virtue of electron mobility.

ELECTRON-TRANSFER REACTION. An electrode surface phenomenon involving an oxidation—reduction reaction, which is generating a faradaic current.

ELECTROOSMOSIS. The phenomenon whereby an electric field moves a fluid through a membrane. Also, charge separation in an electrolyte by osmotic action.

ELECTROSTATICS. A branch of physics dealing with attractions and repulsion between static electric charges.

ELF. Extra low frequency. Electromagnetic energy between 30 and 3,000 Hz.

ELLIPSE (ELLIPSOID) OF ANISOTROPY. In an anisotropic, homogeneous medium, the equipotential surfaces about a point current source which are ellipsoids of revolution flattened by the ratio λ, which is the COEFFICIENT OF ANISOTROPY.

ELTRAN. An early electrical transient exploration concept, similar in principle to more modern time-domain electromagnetic methods, which used a large-spaced dipole—dipole electrode array for petroleum exploration.

EM COUPLING. See COUPLING.

EMU. The electromagnetic system of units, similar to the cgs system except that practical electrical units (except the abampere) are used.

ENHANCEMENT. Improvement of data or interpretation by a filtering or noise-rejection process.

EQUILIBRIUM CONDITIONS. (1) A balanced reaction condition at a state of minimum energy where energy is neither produced nor consumed. (2) A condition predicted by the law of mass action where the velocity of forward and reverse reactions of a reversible process are equal. See REVERSIBLE PROCESS.

EQUIPOTENTIAL LINE METHOD. See EQUIPOTENTIAL SURVEY.

EQUIPOTENTIAL SURVEY. Measurement of potentials about a fixed current electrode (or between line electrodes) where there could also be an infinitely remote current electrode. Similar to the pole—pole array in layout. Also see APPLIED POTENTIAL.

EQUIVALENCE OF CHARGEABILITY AND PERCENT FREQUENCY EFFECT. As a rule of thumb, the Newmont M_{331} chargeability in millivolt seconds per volt is about four times the percent frequency effect measured between frequencies spaced a decade apart.

EQUIVALENT CIRCUIT. An electric circuit consisting of analogous lumped or distributed components to demonstrate certain aspects of a phenomenon, such as the electrical properties of a material.

ERROR. See PROBABLE ERROR and ACCURACY.

EXCHANGE CURRENT. A laboratory and theory term used in electrochemistry where the reversible electric current at an electrode is in equilibrium with an electrolyte.

EXPANDER. An IP and resistivity surveying technique in which the electrode separation interval is successively expanded to achieve a greater depth of exploration. Also called SOUNDING or PROBING. Data from an expander can be interpreted to give an estimate of depth to horizontal layers of material with contrasting physical properties, if such conditions exist.

EXPLORATION LOGGING. Use of more widely spaced in-hole electrodes, perhaps combined with surface electrodes, to expand the exploration effectiveness of a drill hole. Also called DETECTION LOGGING.

EXTRINSIC CONDUCTION. Ionic or electronic conduction in solid, electrically conducting materials; it is due to weakly bonded impurities or defects. Also called structure-sensitive conductivity. Compare INTRINSIC CONDUCTION.

F

FARADAIC. Pertains to the electrochemical electron-transfer reaction as in the oxidation—reduction process in an electrolytic cell.

FARADAIC PATH. Passage of current at an electrode resulting from ion diffusion and electrochemical electron-transfer reaction. See WARBURG IMPEDANCE.

FEEDBACK. Use of part of the output of a system or circuit as a partial input, usually for self-correction or control purposes.

FENCE EFFECT. An IP or resistivity anomaly produced by the presence of a nearby grounded conductor, such as a fence. Types of fences may be indicated on IP maps as wood post fence (WPF), steel post fence (STP), etc.

FICK'S LAWS. The diffusion-rate laws of electrochemistry. The time rate of change of diffusion flux is proportional to the concentration gradient and the activity of the electrolyte.

FIELD. (1) Physically, a region of space influenced by some force-producing agent. (2) Geophysically, the great outdoors where purposeful geophysical surveys are made.

FIELD-EFFECT TRANSISTOR (FET). A semiconducting device that uses as input a transverse electric field to vary its conductance and thus control its current output. Ideally, it is a voltage-controlled current source.

FILTER. A device used in electric circuits to enhance signals or suppress noise of certain frequencies or frequency bands to improve signal-to-noise ratio.

BAND-PASS FILTER. A filter that will pass a signal of a given frequency range (for example, the IP frequencies) and will reject noise levels outside this frequency range.

BUTTERWORTH FILTER. A flat response, low-pass filter designed with a low insertion loss.

DIGITAL FILTER. A means of filtering data numerically in the time domain by summing a series of weighted samples at a series of successive time intervals. It can be the equivalent of frequency filtering.

HIGH-PASS FILTER. A filter that rejects noise below a given frequency.

LOW-PASS FILTER. A filter that rejects noise above a given frequency.

NOTCH FILTER. A filter that rejects a given frequency, for example, 60 Hz.

TCHEBYSCHEFF FILTER. An active band-pass filter with an effective steep skirt or maximum steep rolloff, charac-

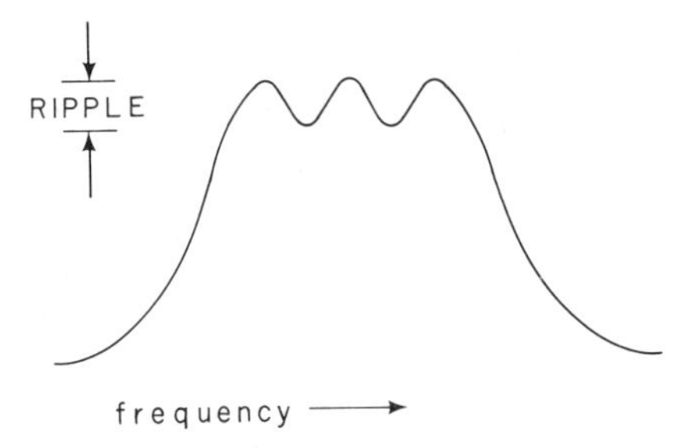

Fig.G.12. A three-pole band-pass Tchebyscheff filter curve with a 1/2 dB ripple. This is then a '1/2 dB Tchebyscheff'. The bandwidth is measured between frequencies below the ripple voltage.

terized by a specified, uniform ripple in the pass band. See Fig.G.12.

FILTERING. Use of a filter to improve signal-to-noise ratio.

LINEAR FILTERING. Convolution, as characterized by impulse response or amplitude and phase response, as a function of frequency.

FIXED LAYER. A compact layer of ions and molecules held rather rigidly in place on an electrode or solid by chemical or electrostatic adsorptive forces. Also called the inner Helmholtz double layer. See DOUBLE LAYER.

FIXED-LAYER CAPACITANCE. Capacitance due to the presence and thickness of the fixed layer.

FLIP-FLOP. A bistable oscillator. A gated circuit used in digital electronics.

FLOATING. Not electrically connected to ground or the system reference voltage.

FLOW CHART. A diagram showing the successive (and alternative) steps in carrying out data processing by digital computer.

FORM FACTOR. (1) The geometric multiplying factor; is dependent on the type of electrode array and interval being used. Also called GEOMETRIC FACTOR. (2) Less commonly, a term used for the type curves found in profiling a subsurface body, for example, a sphere.

FORMATION FACTOR (F). The ratio between rock resistivity ρ_r and pore-water resistivity ρ_w: $F = \rho_r/\rho_w$. The formation factor is approximately proportional to rock porosity squared. See ARCHIE'S LAW.

FORWARD SOLUTION. In mathematical solution of resistivity and IP problems, the precomputation method of predicting electric potentials over a given subsurface structure.

FOURIER ANALYSIS. The analytical representation of any cyclic waveform as a weighted series of sinusoidal functions.

FOURIER TRANSFORMS. Mathematical formulas used to describe a function as an integral or as a series of sinusoidal functions. For instance, a time function can be described as a distribution of frequencies.

FREQUENCY DOMAIN. In circuit theory and measurements, the domain in which frequency is the variable parameter. A term for the frequency method of making IP measurements.

FREQUENCY EFFECT (FE). The difference between voltage or resistivity

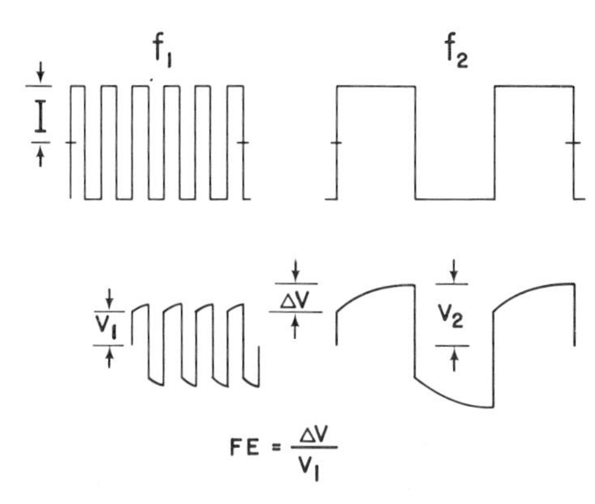

Fig.G.13. Frequency effect (FE).

measurements made at two frequencies divided by the voltage or resistivity at one of the frequencies. Conventionally, FE = $(\rho_2 - \rho_1)/\rho_1$ or more simply, FE = $\Delta\rho_f/\rho_1$. See Fig.G.13.

FULL-WAVE RECTIFIER. A device that uses both positive and negative polarities of alternating current to produce direct current. See Fig.G.8.

FUNDAMENTAL. The base frequency of a periodic function.

G

GAIN. A general term used to denote an increase (or change) in signal power in transmission from one point in a circuit or system to another.

GALVANIC CONTACT. Physical electrical contact of an electrode with the ground as opposed to introduction of electric current by induction. This has been called an ohmic contact if there is no rectification. Also known as a conduction contact.

GATE. An electronic circuit that responds to a certain combination of input signals by giving a particular output signal. Used in digital logic and electronics.

GEOMETRIC FACTOR. A usually constant numerical factor used to multiply the voltage-to-current ratio (resistivity) gained from measurements between electrodes. The geometric factor times resistance gives apparent resistivity. The geometric factor is dependent on the type of electrode array and spacing used. Also called geometric constant, array factor, or sometimes FORM FACTOR.

GISH-ROONEY. (1) An older resistivity surveying method and interpretation. (2) Equipment in which the polarities of both current and potential circuits were alternated (commutated) to cancel effects of polarizing electrodes.

GRADIENT. The rate of change of a spatially dependent function.

GRADIENT ARRAY. See ARRAY.

GROUND. The voltage level of the ground. Also, the reference voltage level of an electrical system or instrument. Ground, of course, refers to the earth itself, especially the soil. As a verb, to connect electrically to ground.

H

HALF-SPACE. A mathematical model of a semispace whose uppermost surface is a plane of infinite extent. In geophysics, the flat earth and its subsurface are usually a close physical approximation.

HARMONIC FUNCTION. Solutions of Laplace's equation. The term is also for periodic functions.

HEAVISIDE FUNCTION. See STEP FUNCTION.

HELMHOLTZ DOUBLE LAYER. See DOUBLE LAYER.

HERTZ (Hz). A unit of frequency replacing cycles-per-second; for example, 60 cps = 60 Hz.

HETEROGENEITY. See INHOMOGENEITY.

HOLE LOGGING. Drill-hole IP or resistivity surveying. See HOLE PROBE.

HOLE PROBE. A drill-hole IP or resistivity survey in which closely spaced, in-hole electrodes are used to determine the electrical properties of rock near the drill hole or are used for near-hole exploration. Also called ELECTRIC LOGGING, IP or RESISTIVITY LOGGING, or simply HOLE LOGGING.

HOMOGENEOUS. Uniformity of physical property regardless of location with a material.

HORIZONTAL PROFILING. See PROFILING.

HYDRODYNAMIC WAVES. Fluid waves that obey dynamic behavioral laws.

HYSTERESIS. A phenomenon exhibited by a system or material in which response to change is nonlinearly dependent on past reactions to change.

I

IDEAL POLARIZED ELECTRODE. In electrochemical theory, a metal-to-electrolyte contact at which no charge crosses the interface. However, charge accumulates, and the electrode interface behaves like a capacitor without leakage. No chemical reaction takes place and there is no exchange current or faradaic process. This condition is approximated when high overvoltage, nonreactive metals are at equilibrium with an electrolyte.

IMAGES, METHOD OF. A means of deriving type curves for use in interpreting resistivity and IP data. Electrode positions are effectively replaced by current sources or sinks whose images substitute for electrical property contrast boundaries.

IMAGINARY. That part of a complex number that is not real. In phase measurements, the out-of-phase component in contrast to the in-phase part. In impedance, the reactive component.

IMPEDANCE (Z). (1) The complex resistance of a circuit, that is, the ratio of driving force to flow. (2) The total opposition to alternating current by an electrical circuit, the magnitude being equal to the square root of the sum of the squares of the resistance and reactance of the circuit. Expressed in OHMS.

IMPEDANCE MATCHING. Equalizing the impedances of two circuits to be connected together to give maximum transfer of power.

IMPULSE. The limit of a single rectangular pulse of unit area as its width approaches zero and its height infinity. An impulse contains all frequencies in equal proportion. Also called Dirac function or delta function.

INACCURACY. See ACCURACY.

INDUCED CURRENT DIPOLE MOMENT PER UNIT VOLUME ($\vec{P}$). A vector parameter describing the fundamental IP properties of a material as a function of chargeability M and current density $\vec{J}$: $\vec{P} = -M\vec{J}$.

INDUCED ELECTRICAL POLARIZATION. See INDUCED POLARIZATION.

INDUCED POLARIZATION (IP). The geophysical phenomenon or exploration method which measures the slow decay of voltage in the ground following the cessation of an excitation current pulse (time-domain method) or low-frequency variations of the impedance of the earth (frequency-domain method). Refers particularly to the phenomenon of electrode polarization (overvoltage) and membrane polarization of the earth and measurements thereof. Also called INDUCED ELECTRICAL POLARIZATION, induced potential, OVERVOLTAGE, or INTERFACIAL POLARIZATION.

INDUCTANCE (L). The capability of an electric circuit or of two adjacent circuits to induce an electromotive force (emf) within the circuit or circuits. Proportional to the ratio between the total flux linking the circuit and the current flow in that circuit. Measured in henrys (H).

INDUCTION. Introduction of current by an alternating magnetic field according to Faraday's law of induction, $e = -(d\phi/dt)$ where e is voltage, ϕ is magnetic flux, and t is time.

INDUCTIVITY. The three-dimensional inductance of a material, measured in henrys per meter. Free space has an inductivity at $1.257 \cdot 10^{-6}$, as do nonmagnetic rocks. Also called PERMEABILITY.

INFINITE ELECTRODE. This is the usually fixed, remote electrode which is at a great distance from the successively moved field surveying electrodes. Its great separation distance ideally has little effect on the measurements of potential differences in the local area of interest.

IN-HOLE IP METHODS. Techniques for measuring near-hole IP and resistivity physical properties using at least one potential and one current electrode in the hole near the point of measurement. See DOWN-THE-HOLE IP METHODS.

INHOMOGENEITY. Lack of uniformity of physical property. Also called HETEROGENEITY.

IN-PHASE. Electrical signal with the same phase as that of the exciting current.

INPUT. The current, voltage, power, or driving force applied to a circuit, system, or device. Also the terminals where these parameters may be applied.

INPUT IMPEDANCE. The ac or dc resistance or impedance across the input terminals of an electrical instrument. If the input impedance of the measuring device is much higher (say, by 100 times) than the circuit impedance to be tested, the measured impedance will not be greatly altered by the presence of the measuring device.

INSULATOR. A nonconductor of electricity.

INTEGRATED CIRCUIT (IC). A single solid-state electrical circuit element replacing sometimes a large number of individual components. Complete amplifiers, oscillators, and other modules are readily available as integrated circuits using microminiature components.

INTEGRATION (OF CHARGEABILITY). Measurement of the area under the IP decay curve by integration of the decay voltage V_t with time. This measurement normalized by dividing by the primary voltage represents one definition of chargeability. The areas under several successive decay curves can be averaged to improve the precision of the measurement.

INTERFACIAL POLARIZATION. A dielectric property due to conductivity inhomogeneities in a material. See INDUCED POLARIZATION.

INTRINSIC CONDUCTION. Conduction due to major components of the material as opposed to conduction due to impurities or imperfections. At high temperatures, intrinsic conduction will dominate the other conduction modes.

INTRINSIC IP. The true IP property of a material, in contrast with its apparent value.

INVERSE PROBLEM. The general problem in interpreting geophysical field data, that is, how to gain knowledge of the physical features of a disturbing body by making observations of its fields and potentials. It is the inverse of the forward method of computing fields from assumed shapes, and it is mathematically more difficult and often nonunique.

ION EXCHANGE. The property of some minerals, particularly clays, of sorbing certain anions and cations and retaining them in a state whereby they can be exchanged for other anions and cations in solution. Ion exchange is a diffusion process, and its rate depends on ion mobility.

ION MOBILITY (μ). Ease of movement of ions in an electric field. Measured by the ratio of ion velocity to electric field strength.

IP. See INDUCED POLARIZATION.

IP SUSCEPTIBILITY. One term for the measure of the induced-polarization property of the earth. Implies an analogy with other types of polarization properties of earth materials, that is, induced magnetic effects (Bleil, 1953).

ISOTROPIC. Having the same physical property in all directions; that is, system equation coefficients are scalar.

K

k. See RESISTIVITY CONTRAST FACTOR.

KERNEL. A function of apparent resistivity and variable to be integrated; used in the interpretation of resistivity data (Slichter, 1933).

K INDEX. A measure of the average intensity of magnetic disturbance with time. Includes magnetic storms but not diurnal and lunar variations.

KINETIC. In electrochemistry, refers to rate processes and the movement of ions and electric current in a solution and at an interface.

KIRCHOFF'S LAWS. In any electrical circuit:

(1) The algebraic sum of currents flowing toward any point or node is zero.

(2) The algebraic sum of electromotive forces and products of current and resistance around any closed circuit are equal. Or, the sum of all voltages around any closed path in a circuit is zero. Kirchoff's laws can also be applied to three-dimensional materials, such as the earth, to assist in the interpretation of apparent-resistivity curves.

L

L. A symbol for the area over a decay curve, extending from 0.45 to 1.75 seconds. See COMPLEMENT OF CHARGEABILITY.

LAPLACE'S EQUATION. The relationship showing that the Laplacian ∇^2 of a potential function V vanishes in space which contains neither sources or sinks, or $\nabla \cdot \nabla V = 0$. ∇^2 is the Laplacian, a vector operator. In rectangular coordinates:

$$\nabla^2 V = \frac{\partial^2 V}{\partial x^2} + \frac{\partial^2 V}{\partial y^2} + \frac{\partial^2 V}{\partial z^2} = 0$$

LATERAL. (1) A well-logging electrode array similar to the pole—dipole array in which the infinite current electrode is on the surface. (2) Sideways.

LINEAR. Refers to the straight-line relationship displayed if plotting a constant of proportionality, such as resistance R in Ohm's law, $V = IR$, where V and I are variables.

LINEAR FILTER. Convolution. A mathematical operation using impulse responses.

LINEARITY OF INDUCED POLARIZATION. (1) Refers to the proportional relationship between polarization $\vec{P}$ and current density $\vec{J}$, in which chargeability M is a constant. (2) Refers to the symmetrical identity between voltage and time of the IP charge and decay curves. See LINEAR SYSTEM.

LINEAR SYSTEM. (1) A system whose response or output is linearly related to its input. If a linear system is excited by an input sine wave of frequency f_1, the output will contain only the frequency f_1; however, the output amplitude and phase may be changed. (2) An electrical circuit whose impedance is independent of the applied current or voltage.

LOG. A record of one or more physical measurements as a function of depth in a borehole.

LOGARITHMIC CONTOUR INTERVAL. Plotting of data on a logarithmic scale, which is sometimes used where properties of materials vary by many orders of magnitude. It is frequently more convenient to plot and contour resistivity and IP data in logarithmic (geometric) intervals rather than in equi-incremental arithmetic intervals. For example, 1, 2, 5, 10, 20, 50, 100 or 1, 3, 10, 30, 100, 300, etc.

LOGGING. Measurements of near-hole electrical physical properties. See HOLE LOGGING.

LOGNORMAL DISTRIBUTION. A statistical distribution in which data, if plotted logarithmically, have the appearance of a normal Gaussian distribution.

LOSS TANGENT (δ). A measure of dielectric loss defined by the relationship tan $\delta = \sigma/(\epsilon\omega)$ where σ is conductivity, ϵ is dielectric permittivity, and ω is angular frequency.

LUMPED. Refers to electrical network analysis, the collection or lumping of resistance, capacitance, or other elements into discrete circuit elements. See DISTRIBUTED.

M

M. See CHARGEABILITY.

MAGNETIC INDUCED-POLARIZATION METHOD (MIP). Measurement of the IP phenomenon by the sensing of magnetic field variations near an electrically polarized subsurface body. Conventionally, the current is introduced into the ground through current electrodes in an arrangement similar to the gradient array and magnetic field measurements are made by means of a sensitive detecting element.

MAGNETIC STORM. Rapid, irregular, transient fluctuations of the magnetic field, which are greater in magnitude, more irregular, and more rapid than diurnal variations and occur most commonly during unusual sunspot activity.

MAGNETOHYDRODYNAMICS. The branch of physics that deals with phenomena associated with the motion of an electrically conducting fluid, as a liquid, metal, or ionized gas, through a magnetic field. Also called hydromagnetics.

MAGNETOSPHERE. The space from the earth outward that is pervaded by the earth's magnetic field. Extends to more than 10 earth radii.

MAGNETOTELLURIC NOISE. Unwanted voltage signals due to low-frequency earth currents whose source is electrical discharges in thunderstorms, power lines, ionospheric currents, or magnetospheric currents.

MASSIVE SULFIDES. Rocks that contain more than 20% sulfides by volume. Is also sometimes used to refer to electrical phenomena from materials that behave like massive metallic substances. Compare DISSEMINATED SULFIDES.

MASTER CURVES. See TYPE CURVES.

MEMBRANE POLARIZATION. The weaker IP effect due primarily to restrictions of ion mobility as opposed to electrode polarization. See also NORMAL EFFECT.

MESSAGE. The desired information being sought, usually from field data, in contrast with signal, which is the summation of message and noise.

METAL FACTOR (MF). A parameter sometimes used in the interpretation of IP data whereby percent frequency effect (PFC) or chargeability (M) is normalized by dividing by the measured resistivity and multiplying this ratio by a constant to put it in the range of conveniently used numbers. The metal factor measures the total change in conductivity or the capacitivity of a rock. When ρ is in ohm-feet:

$$MF = \frac{PFE}{\rho/(2\pi)} \cdot 10^3$$

(1) Originally, the metal factor was defined as:

$$MF = 2\pi \frac{\rho_{dc} - \rho_{ac}}{\rho_{ac}\rho_{dc}} \cdot 10^5$$

where ρ_{dc} is the low-frequency resistivity and ρ_{ac} is the high-frequency resistivity in ohm-feet.
(2) Now most groups working in the frequency domain define:

$$MF = 2\pi \frac{\rho_{dc} - \rho_{ac}}{\rho_{ac}^2} \cdot 10^5$$

because of instrument measuring techniques.
(3) In the time domain, the metal factor is $(M/\rho_{dc}) \times 2{,}000$ where M is chargeability in millivolt-seconds per volt and ρ_{dc} is in ohm-meters. In the time domain, this unit is similar to the parameter defined by Keller (1959) as SPECIFIC CAPACITY, measured in farads per meter.
(4) Metal factor is proportional to electric susceptibility (χ), a dielectric property that is a constant of proportionality between induced current dipole moment per unit volume ($\vec{P}$) and electric field ($\vec{E}$) in the relation $\vec{P} = -\chi\vec{E}$:

$$MF = \chi \cdot 2 \cdot 10^6$$

The metal factor has units of conductivity or capacitivity and is also called METALLIC CONDUCTION FACTOR (MCF).

METALLIC CONDUCTION FACTOR (MCF). See METAL FACTOR.

MHO. The unit of conductance or admittance. The reciprocal of the OHM.

MHO PER METER. The unit of conductivity. Reciprocal of ohm-meter.

MICROPULSATION. Relatively rapid, oscillatory, small magnetic disturbances in the frequency range 0.01—3 Hz. Micropulsations are classified into continuous (Pc), irregular (Pi), pearl (PP), and other types, depending on their character.

MIGRATION (OF IONS). The motion of ions in a solution as a result of an electric field gradient. This being an electrostatic process, the driving force does not involve solution concentration gradients (diffusion) or a faradaic process. See MOBILITY.

MILLIRADIAN. A unit of phase measurement equal to 0.0573 degrees of arc. One degree equals 17.45 milliradians.

MILLISECOND. A standard field measurement unit of chargeability, measured as the area under the decay curve of a pulsed (+, 0, —, 0) square wave. See CHARGEABILITY. Also, of course, a unit of time.

MIP. See MAGNETIC INDUCED-POLARIZATION METHOD.

MISE-À-LA-MASSE. A drill-hole IP or resistivity survey technique whereby a buried anomalous body is energized by putting a current electrode in it with potentials being measured on the surface or elsewhere hopefully showing the projected trend direction of the body.

mks. The meter-kilogram-second system of units.

MOBILITY (OF CHARGE CARRIERS). The velocity of charge carriers per unit electric field. See MIGRATION OF IONS.

MODEL. A concept from which effects can be deduced and compared to observations to assist in developing an understanding of the observations. The model may be conceptual, physical, or mathematical.

MULTIPLIER. See ATTENUATOR.

MUTUAL. Refers to inductance, capacitance, or resistance (impedance) coupling between circuits, as in the transmitter and receiver circuits of an IP survey system. See COUPLING.

N

NEAR-DC. A descriptive term for the commutated dc or low-frequency ac current used in resistivity and IP surveying to emphasize the propagation characteristics of electricity.

NEGATIVE IP EFFECT. An IP decay voltage opposite in sign to that of the charging current and due to the geometric relationship of a shallow polarizable body and the measuring electrode array.

NOISE. Any unwanted voltage not due to the desired effect being measured. Noise can be due to atmospheric electrical discharges (sferics), 60-Hz interference from power lines, motor generator or electronic components, or low-frequency magnetotelluric phenomena (telluric noise). Geologic noise (Slichter, 1955) is interference from extraneous geological conditions.

NONFARADAIC PATH. The path for the charging of the double layer by the virtual passage of a current near an electrode as a result of reorientation of the ionic layers of the double layer. The process is analogous to charging a capacitor and no net charge is transported across the interface.

NONPOLARIZABLE ELECTRODE. An electrode whose potential is not affected by the passage of a current through it. See POROUS-POT ELECTRODE.

NORMAL ARRAY. A well-logging electrode array similar to the pole—pole array in which the infinite current and potential electrodes are outside the hole. A spacing of about 16 inches is used for the short normal and 64 inches for the medium or long normal.

NORMAL EFFECT. An unwanted background IP effect due in part to membrane polarization; found to some extent in most rocks. See BACKGROUND POLARIZATION.

NORMALIZATION. The method of using the ratio between the observed data and a particular standard (normal) datum to provide a convenient scale factor for a model presentation. A normalized value usually does not have units and is sometimes given in percentage.

NORMALIZED APPARENT RESISTIVITY. The ratio between apparent resistivity and the resistivity of the upper layer. In constructing type curves, normalized apparent resistivity is plotted against normalized electrode interval (the ratio between the electrode interval and the thickness of the upper layer). See APPARENT RESISTIVITY CURVE.

n-TYPE SEMICONDUCTOR. A doped semiconductor with more electrons than holes available for carrying charge.

O

OCTAVE. A difference in frequency of a factor of 2 or 1/2. Filter rolloff is sometimes specified in terms of decibels per octave.

OFF TIME. The time during which an IP pulse-type transmitter is off and the decay voltage is measured at the receiver.

OHM. The unit of electrical resistance.

OHMIC. See LINEAR or GALVANIC CONTACT.

OHM'S LAW. The common situation in which the current I in a circuit is directly proportional to the electromotive force V in the circuit. Expressed algebraically as $V = IR$ where R is resistance in ohms. Rocks do not necessarily obey Ohm's law at high current densities.

ON TIME. The time during which an IP transmitter is actually supplying current. The time during which the charging current from a pulse-type transmitter is received at the receiver.

OPERATIONAL AMPLIFIER. A high-gain, high-input impedance amplifier requiring minimal current for operation. Ideally, a voltage-controlled voltage source.

ORDER OF MAGNITUDE. An interval of ten, or multiples of ten, more or less than a given value. The term is sometimes used to give the error or uncertainty of measurement if the uncertainty is large.

OUT-OF-PHASE. That component of an electrical signal that has a 90-degree phase difference from the exciting or reference current. Also called QUADRATURE.

OUTPUT. The power, current, or voltage delivered by a circuit, system, or device. Also the terminals where the power, current, or voltage may be delivered.

OVERVOLTAGE. The extra potential, which in IP is proportional to impressed current density, due to an electrochemical and electrokinetic barrier set up at an electrode—electrolyte interface. Activation overvoltage is caused by passage of current stimulating an electron-transfer reaction such that the electrode deviates from its reversible potential without appreciably changing ion concentrations at the electrode surface. Concentration overvoltage is brought about by a depletion or accumulation of oxidized and reduced ion species at the electrode surface causing a change in the reversible potential of the electrode. See INDUCED POLARIZATION.

P

Pc. Continuous-type micropulsation, which can be subdivided into classes according to period or frequency.

Pi. Irregular-type micropulsations.

PARADOX OF ANISOTROPY. The finding that, contrary to what would be expected from the true resistivity, the apparent resistivity of steeply dipping foliated rocks measured by an array in a direction normal to bedding is less than that measured parallel to the bedding. It is due to the greater current density along the plane of stratification. See COEFFICIENT OF ANISOTROPY.

PARALLEL FIELD. A uniform electric field, that is, one in which current flow lines and potential surfaces are parallel. See PLANE-WAVE FIELD.

PASSIVE SYSTEM. A system or circuit with no internal source of energy. Also, a system or circuit that does not generate an output if there is no input. Compare ACTIVE SYSTEM.

PDMI. See PERCENT DECREASE IN MUTUAL IMPEDANCE.

PDR. See POTENTIAL DROP RATIO.

PERCENT DECREASE IN MUTUAL IMPEDANCE (PDMI). The percent voltage change of a coupled circuit with respect to the low frequency. Used with reference to the amount of in-phase electromagnetic coupling.

PERCENT FREQUENCY EFFECT (PFE). The basic polarization parameter measured in the frequency domain, that is, the percent difference in resistivity measured at two frequencies. It is simply the ratio, $\Delta\rho_f/\rho$, measured in percent, where $\Delta\rho_f$ is the IP response observed as a frequency-dependent resistivity difference and ρ is resistivity. See Fig. G.10.

(1) By accepted definition (Madden and Marshall, 1958):

$$\mathrm{PFE} = \frac{\rho_{dc} - \rho_{ac}}{\rho_{ac}} \times 100$$

where ρ_{dc} and ρ_{ac} are the low- and high-frequency resistivities, respectively.

(2) A standardization of decade normalization has been proposed whereby normalized PFE is:

$$\mathrm{PFE} = \frac{\rho_{dc} - \rho_{ac}}{\rho_{ac}} \; \frac{1}{\log_{10} \dfrac{f_{ac}}{f_{dc}}} \times 100$$

(3) Keller (1966) has suggested that PFE be defined as:

$$\mathrm{PFE} = \frac{\rho_1 - \rho_2}{(\rho_1\rho_2)^{1/2}} \times 100$$

where ρ_1 and ρ_2 are resistivities measured at a decade frequency change.

(4) PFE is closely related to chargeability if defined as suggested by Brant (1964) as:

$$\mathrm{PFE} = \frac{\rho_{dc} - \rho_{ac}}{\rho_{dc}} \times 100$$

PERCENT MINERALIZATION. In IP surveying, this usually refers to the percent by volume of metallic-luster minerals in a rock. This value is usually about half the metallic-luster mineral content by weight.

PERIOD. The time T for one cycle. The time interval between two successive wave points. $T = 1/f$, where f = frequency.

PERMEABILITY. See INDUCTIVITY.

PERMITTIVITY (K). The name for the

volumetric capacitance of capacitivity of a three-dimensional material, such as a dielectric. The permittivity of free space is $8.85 \cdot 10^{-12}$ F/m. See DIELECTRIC CONSTANT. Relative permittivity is the dimensionless ratio between the permittivity of a material and that of free space.

PFE. See PERCENT FREQUENCY EFFECT.

PHASE. The angle of lag or lead (or the displacement) of a sine wave with respect to a reference. Commonly expressed in angular measure. Consists of an in-phase (real) component and an out-of-phase (imaginary) part.

PHASE ANGLE. (1) In induced polarization, descriptive of a method of IP surveying or a measurement of IP effect. (2) Phase angle β is defined as:

$$\beta \approx \tan^{-1}\left(\frac{\rho_{\text{out-of-phase}}}{\rho_{\text{in-phase}}}\right)$$

In induced polarization, phase angle is usually measured in milliradians. (3) Phase-angle response can be separated into real (in-phase) and imaginary (out-of-phase) components and can be plotted as such on a Cole—Cole plot.

PHASE CONTROL. The process of rapid on—off switching which connects an ac supply to a load for a controlled fraction of each cycle. An efficient use of silicon controlled rectifiers in regulating current in an IP transmitter.

PHASE LOCK. A type of synchronous detection in which an internal constant or almost constant frequency signal is generated within the detecting instrument, and this is brought to the same average phase as the external signal. Used for synchronous detection of the external signal to suppress noise.

PHASOR DIAGRAM. A diagrammatic display of a vector which rotates at some particular angular velocity. By viewing the 'frozen' position and spatial relationship of the vector, characteristic behaviors can be interpreted.

PLANE-WAVE FIELD. A uniform electric or electromagnetic field in which the potential surfaces are planar. See PARALLEL FIELD.

PLOT POINT. The mapped point where a value is plotted relative to a point on the measuring electrode array; for a symmetric array it is the array midpoint, but on an asymmetric array (no plane of symmetry) the plot point may vary according to the policy of the surveying group.

PLOTTED PSEUDOSECTION. A method of displaying IP and resistivity data for review and interpretation by plotting profile values horizontally and expander values vertically and contouring the resulting patterns. Called a pseudosection because it cannot be interpreted as directly representing earth properties in section. Sometimes pairs of resistivity and IP pseudosections are plotted as mirror images, the upper section being inverted. See Fig.G.2.

POINT SOURCE. A single current electrode whose companion is a great distance away. This is the current pole of the pole—pole or pole—dipole arrays.

POLARIZATION. (1) Dipole moment per unit volume. (2) In induced polarization, the production of current dipole moment per unit volume. (3) Also in IP work, refers to polarity or potential near an electrode. See INDUCED POLARIZATION.

POLARIZATION RESISTANCE. The factor $[RT/nF\vec{J}_0]$ as employed in overvoltage theory, relating overvoltage η to current density $\vec{J}$ by:

$$\eta = -\vec{J}\left(\frac{RT}{nFJ_0}\right)$$

where R is the gas constant, T is absolute temperature, n is the number of molar equivalents, F is the faraday, and J_0 is exchange current density, all in cgs units. The units of polarization resistance are ohm-cm^2 for an electrode one cm^2 in area.

POLARIZED ELECTRODE. See ELECTRODE POLARIZATION and IDEAL POLARIZED ELECTRODE.

POLE. The single electrode of a pair from which IP measurements may be made. Its companion is called the infinite pole or infinite electrode if it is far enough away so that its presence does not affect the measurements.

POLE—DIPOLE ARRAY. See ARRAY.

POROUS-POT ELECTRODE. A type of nonpolarizable electrode through which negligible current flows used as a ground contact in making voltage measurements in field resistivity and IP surveying. A copper electrode is immersed in a copper sulfate solution contained in an unglazed porous porcelain pot. Also called a pot. See NON-POLARIZABLE ELECTRODE.

POT. See POROUS-POT ELECTRODE.

POTENTIAL. The work done when a unit charge, pole, or mass is brought from infinity to a point. Electrical potential is measured in volts.

POTENTIAL DIFFERENCE. The algebraic difference between individual potentials at two points. Measured by the work done in moving a unit charge from one point to the other. Measured in volts.

POTENTIAL DROP RATIO (PDR). An electrical survey method which compares ratios of voltages between two adjacent aligned sets (pairs) of potential electrodes.

POTENTIAL ELECTRODE. A contact of the IP or resistivity receiver circuit with the ground, usually a porous-pot electrode.

POTENTIAL GRADIENT. At a point, the potential difference per unit length, usually measured in the direction in which it is a maximum. Units are volts per unit length. Also see ELECTRIC FIELD.

POTENTIOMETER. An electrical instrument for measuring low-level, zero-frequency or dc voltages without drawing current from the measured circuit by using the unknown voltage as an arm in a dc bridge circuit.

POT RESISTANCE. The electrical resistance at a porous-pot potential electrode due to the resistance of adjacent material. Too high a pot resistance will 'load' the voltmeter-receiver's input impedance, reducing the reading and making it susceptible to electrical noise.

POWER. The rate of doing work, measured in watts (J/s). The product volt-amperes or force-times-flow.

POWER SPECTRUM. A graph of power density (spectral density or power spectral density) against frequency. The power spectrum is the square of amplitude plotted against frequency.

PREAMPLIFIER. An amplifier in front of the main amplifier. Located nearest the signal source to improve signal-to-noise ratio. Often has a high input impedance to prevent loading and to give maximum signal transfer.

PRECISION. The repeatability of an instrument. Does not cover other sources of error which may be inherent or present in the measuring system.

PRIMARY VOLTAGE (V_p). In IP surveying, the peak asymptotic charging voltage observed at a time-domain receiver. This occurs during the later time that current I_p is being put into the ground at the IP transmitter. Used to determine apparent resistivity and chargeability. See Fig. G.10.

PRINTED-CIRCUIT (PC) BOARD. A thin glass laminate or plastic composition board on which electrical circuits are drawn and components mounted. Can be easily removed for testing and replacement.

PROBABLE ERROR. The range of values within which half of a series of readings will probably lie. For a Gaussian distribution, the probable error is 0.648 times the standard deviation.

PROBING. See SOUNDING.

PROFILING. A resistivity or IP field surveying method in which an array with a fixed electrode interval is moved progressively along a traverse to create a horizontal profile of the apparent physical property of the section. Usually refers to horizontal profiling but may conceivably refer to vertical profiling, or sounding.

PROPAGATION CONSTANT (k). In electromagnetic theory, the relation $k^2 = \mu K \omega^2 + i\mu\sigma$ where μ is magnetic permeability, K is dielectric permittivity, ω is angular frequency, and σ is electrical conductivity. Also called WAVE NUMBER.

PSEUDOSECTION. See PLOTTED PSEUDOSECTION.

p-TYPE SEMICONDUCTOR. A doped semiconductor with more holes than electrons available for carrying charge.

PULSE CURVE. See BETA CURVE.

PULSE IP METHOD. A name for the time-domain IP method in which a transmitted signal may cause a voltage decay phenomenon after cessation of the transmitter current pulse. Also called the pulse potential, the dc pulse, or simply pulse method.

Q

QUADRATURE. See OUT-OF-PHASE and PHASE ANGLE.

QUADRATURE CONDUCTANCE. See METAL FACTOR.

QUASI-SECTION. See PLOTTED PSEUDOSECTION.

R

RADIAL SURVEY. See AZIMUTHAL SURVEY.

RANDOM. A value not in a consistent pattern but within phenomenologic limits.

REACTANCE. That part of electrical impedance that is due either to capacitance or inductance or both, expressed in ohms. The imaginary component of impedance.

REACTION, REVERSIBLE. See REVERSIBLE PROCESS.

REAL. The in-phase component of a complex number.

RECEIVER. As used in IP field surveys, a sensitive, filterable, ac or dc voltmeter with SP buckout controls. Generally, but not always, the frequency-domain receiver is ac coupled and the time-domain voltmeter is dc coupled.

RECIPROCITY, PRINCIPLE OF. In potential theory as applied to electrical exploration, the principle that the potential at a point M with respect to a current source at A is the same as if the points of measurement were reversed.

RECTIFIER. A device that passes current

in only one direction. Used for converting an alternating current into a unidirectional current by the inversion or suppression of alternate half-waves.

REFLECTION COEFFICIENT (k). A ratio of differences of resistivity as derived from the method of images:

$$k = \frac{\rho_2 - \rho_1}{\rho_2 + \rho_1}$$

Used in describing type curves. Also called RESISTIVITY CONTRAST FACTOR or REFLECTION FACTOR.

RELATIVE APPARENT RESISTIVITY (ρ_a/ρ). The ratio between the apparent resistivity and the true resistivity of a portion of a model; for example, apparent resistivity divided by the true resistivity of the upper layer in a simple two-layer case. Used as a dimensionless ratio or scale factor in the presentation of resistivity type curves.

RELATIVE PERMITTIVITY. Dielectric constant (K) normalized by dividing by the permittivity of free space ($8.25 \cdot 10^{-12}$ F/m) so as to give a dimensionless quantity.

RELATIVE THICKNESS (n/a). The ratio between the thickness n of a layer and the electrode interval a. Expressed as a dimensionless ratio in the presentation of resistivity data or resistivity models, such as normalized apparent resistivity curves.

RELAXATION TIME. See TIME CONSTANT.

RELAY. A switching device, usually controlled by a separate electrical circuit. Mercury-wetted relays are sometimes used to switch current in IP transmitters.

REMOTE ELECTRODE. The distant (or infinite) current electrode that is the companion to the current source about which an IP or resistivity survey is made.

REMOTE SENSING. A method of using natural or artificial fields to detect contrast in a physical property or geometric position without making physical contact. An active sensor requires a source of energy.

REMOTE TRIGGERING. A method of recording a voltage or decay signal by controlling the on and off voltage switching of the receiving equipment. A synchronous detection method that uses the ground signal for a timing channel.

REPEATABILITY. A maximum deviation from the average of corresponding data points taken from repeated tests under ideal conditions.

RESISTANCE (R). An electric property of a circuit or of a circuit element linearly opposing the passage of current. According to Ohm's law, it is equal to the ratio of voltage V to current I across the circuit: $R = V/I$, where R is in ohms. Also called ohmic resistance. Also descriptive of the circuit element displaying the property of resistance.

RESISTIVITY (ρ). The low-frequency electrical property of a three-dimensional substance that is a measure of the difficulty in driving an electrical current through it, properly measured in ohm-meters. Resistivity is the reciprocal of conductivity σ: $\rho = 1/\sigma$. If the material is anisotropic, ρ and σ are tensors and σ^{-1} is interpreted as a matrix inversion operation on ρ. Also called specific resistance. Some include the half-space dimensional geometric factor 2π with resistivity in ohm-feet, defining unit resistivity as $1.0\ \rho/2$ ohm-feet = 1.9 ohm-meters. See APPARENT RESISTIVITY.

RESISTIVITY CONTRAST FACTOR (k). See REFLECTION COEFFICIENT.

RESISTIVITY INDEX (ρ_1/ρ_2). The true (not apparent) resistivity ratio of a model. Used as a dimensionless ratio, or scale factor, in the presentation of type curves.

RESISTIVITY METHOD. A method of studying the earth resistivity by observing the electric fields caused by the introduction of low-frequency current into the ground used in geophysical exploration.

RESISTIVITY SPECTRUM. A recording of the ac resistivity of a polarizable material measured at successive frequencies and plotted against frequency. This spectrum can be transformed into an IP decay curve and vice versa. The IP resistivity spectrum of a polarizable material often appears to be indicative of the nature of the type of mineralization of the substance. Also called CONDUCTIVITY SPECTRUM.

RESOLUTION. An expression of the ability to detect a body and perhaps measure its size or to indicate clearly each of the anomalies due to two adjacent structures. The resolution of resistivity or IP equipment is related to electrode interval, property contrast, background values, and geometric relationships of the body or bodies to the electrodes.

REVERSIBLE PROCESS. A physical or chemical change that can be caused to proceed in either direction by an infinitesimally small alteration in one of the conditions controlling equilibrium, such as concentration, temperature, or pressure. If a small change in one condition shifts the equilibrium position, a change back to the original value restores the equilibrium position to its original position. See EQUILIBRIUM CONDITIONS.

rms. See ROOT-MEAN-SQUARE.

ROLLOFF. The attenuation of a filtered signal with frequency, often given in decibels per octave.

ROOT-MEAN-SQUARE (rms). The square root of the arithmetical mean of the squares of a set of numbers. The root-mean-square of the instantaneous values over one cycle is used in electrical measurements because it is related to the power dissipated by a resistor.

S

SATURATION. (1) An apparent nonlinear resistivity or IP behavior due to large contrasts in electrical property or extreme values of the resistivity contrast coefficient. Under such conditions, it may be difficult to evaluate the true resistivity or IP effect of a body but relatively easy to find its depth because of what is called the resistivity saturation effect. In IP measurements, the IP response sometimes varies nonlinearly with charging current, which has been called an IP saturation effect and is probably due to exceeding the current density limit of the polarizable body. (2) The limiting value of a nonlinear variable.

SATURATION PROSPECTING. The blanket use of several exploration methods on an area in contrast to a more selective choice of exploration methods.

SCHLUMBERGER ARRAY. See ARRAY.

SCHUMANN RESONANCE. An electromagnetic wave-guide phenomenon between the earth and the ionosphere in which characteristic extra-low frequencies (ELF) are propagated.

SCR. Silicon controlled rectifier. See CONTROLLED RECTIFIER.

SECONDARY VOLTAGE (V_s). In IP surveying, the polarization voltage observed at a time-domain receiver immediately after the primary current is turned off. Also sometimes called initial transient voltage or initial decay voltage.

SEG. Society of Exploration Geophysicists.

SELF-POTENTIAL (SP). The effect of the dc and slowly varying natural ground voltages observed between nearby nonpolarizing electrodes in field surveying. In many mineralized areas, self-potential is caused by a double half-cell electrochemical reaction at an electrically conducting sulfide body. Also called SPONTANEOUS POLARIZATION.

SELF-RESISTANCE (SELF-IMPEDANCE). The resistance (impedance) of a circuit element. The term can be applied to the resistance of a single electrode. See MUTUAL RESISTANCE.

SEMICONDUCTOR. A substance, such as germanium or silicon, whose electrical conductivity at normal temperature is usually intermediate between that of a metal and an insulator. Its concentration of charge carriers increases with temperature over a given range. Many common metallic sulfides and oxides are actually semiconductors.

SENDER. The IP current waveform generator—control device used in the field. Also called TRANSMITTER.

SENSITIVITY. The least change in a quantity that a detector is able to perceive. An instrument can have excellent sensitivity and poor accuracy.

SFERICS. Short for atmospherics, natural fluctuations of the electromagnetic field, generally at frequencies from 1 to 100,000 Hz, caused principally by lightning discharges in the earth's atmosphere. See SCHUMANN RESONANCE.

SHIELDING. Refers to the enclosing of the electrode wires or components commonly found in an IP receiver to reduce the effects of noise and electrostatic, magnetic, or electromagnetic coupling.

SIGNAL. The sought-for voltages that carry information pertaining to an electrical measurement as opposed to noise.

SIGNAL-TO-NOISE RATIO. A ratio of wanted to unwanted signals used as a measure of the efficiency of a method or a circuit.

SIGNATURE. A characteristic, interpretable change in a signal due to an anomalous condition.

SKIN DEPTH (δ). The effective depth of penetration of electromagnetic energy. The depth at which plane-wave ac currents penetrating in a conductor will be diminished to $1/e$ (37%) of their original surface density:

$$\delta = \sqrt{\frac{1}{\omega \mu \sigma}}$$

where δ is the skin depth in meters, σ is conductivity in mhos per meter, ω is angular frequency in radians per second, and μ is permeability in henrys per meter.

SKIN EFFECT. An electromagnetic current propagation phenomenon by which high-frequency alternating current flows on the surface of a conductor because of destructive eddy current interference at depth.

SOLAR WIND. High-velocity ionized gas flowing radially outward from the sun. Transient magnetic disturbances as well as the K index may be correlated with variations in the solar wind.

SOLID-STATE CIRCUITRY. The liberal use of semiconductors, such as transistors, integrated circuits, etc., in circuits. These devices do not require much space or power.

SORPTION. The binding of one substance to another by any mechanism, such as adsorption or absorption.

SOUNDING. A type of resistivity or IP survey in which the electrode interval is successively increased relative to a fixed point in the electrode array. Can give electrical information at increasing depths and can be interpreted to show the depth and property contrast of horizontal layers, if such exist, in the survey area. Also called EXPANDER or PROBING.

SP. See SELF-POTENTIAL.

SP BUCKOUT. A variable voltage compensation circuit in series with the input terminals of an IP or resistivity or SP receiver. Used to match the input voltage level of the voltmeter-receiver to that of the ground. The buckout voltage is equal and opposite to the dc self-potential. Also called SELF-POTENTIAL COMPENSATION.

SPECIFIC CAPACITY. A polarization parameter in the time domain similar to the metal factor in the frequency domain, noted by Keller (1959). It is the long-time chargeability (area under decay curve) divided by resistivity. Also called STATIC CAPACITY or CAPACITIVITY. Dimensions are farads per meter.

SPECIFIC IMPEDANCE. See RESISTIVITY. 'Specific' refers to the normalized dimensional or volume properties of a material.

SPONTANEOUS POLARIZATION. See SELF-POTENTIAL.

SQUARE WAVE. A full square wave is a current (or voltage) waveform consisting of equal magnitude positive on and negative on portions. A half square wave is switched on and off. A pulsed square wave has positive on and off and negative on and off portions as are generated by an IP pulse transmitter.

STACKING. As used in time domain, a method of reducing noise voltages by summing (averaging) the decay voltages of several decay cycles.

STAKE RESISTANCE. The electrical resistance at a current stake electrode due to material adjacent to that with which it is in contact.

STANDARD DEVIATION. The square root of the arithmetical mean of the squares of the deviations from a mean value.

STATIC CAPACITY. See SPECIFIC CAPACITY.

STEADY STATE. Equilibrium conditions observed when short time variations are absent.

STEP FUNCTION. A single change in level of a constant, straight-line representation. The on or off switching of current is a step function that can be mathematically translated into a periodic frequency function. The first derivative of a step function is an impulse. Also called HEAVISIDE FUNCTION.

STREAMING POTENTIAL. A weak potential or voltage generated when an ionic fluid passes through a porous solid. This electrokinetic phenomenon is due to a zeta potential and is sometimes seen in measuring self-potential, especially in electric well logging.

SUPERPOSITION, PRINCIPLE OF. The convenient principle that potentials can be combined algebraically, that is, the potential at a point due to a nearby current electrode pair is the same as though a current I_1 emanates from the first electrode and a current I_2 emanates independently from the second electrode.

SURFACE CONDUCTIVITY. Conduction along the surfaces of certain mineral grains due to excess ions in the diffuse layer.

SURFACE POLARIZATION. Polarization at the surface of a body.

SYNCHRONOUS DETECTION. A method of enhancing signal and suppressing noise in detecting waveforms by synchronizing the detection period of the voltmeter-receiver with the on cycle of the current transmitter. Similar to PHASE LOCK.

T

TAFEL'S LAW. A theoretical relationship between overvoltage η and current density $\vec{J}$ at an anode or cathode:

$$\eta = a - b \log_{10} \vec{J}$$

where a and b are experimentally determined thermodynamic constants. This law applies over a greater current density range than is used in IP field measurements.

TAGG'S METHOD. A method of interpreting the resistivity sounding data obtained with the Wenner array over a layered earth.

TCHEBYSCHEFF FILTER. See under FILTER.

TELEMETERING. The transmission of data from a point of observation to a recording point, usually by radio. Telemetered time signals can be used for synchronous detection of resistivity and IP signals.

TELLURIC CURRENTS. Earth currents of very low or zero frequency extending over large regions, which may vary cyclically in direction. Telluric currents are widespread, originating in variations of the earth's magnetic field, as opposed to local currents generated by electrochemical action in mineralized regions.

TELLURIC METHOD. Use of voltage gradients developed in the earth by naturally flowing electric currents to study variations in earth resistivity.

TELLURIC NOISE. See NOISE.

TENSOR. A quantity or set of functions that relates spatially different vector fields, such as may be involved in a change in coordinate system. For example, electric field $\vec{E}$ and current density $\vec{J}$ are related by resistivity ρ in which $\vec{E} = \rho\vec{J}$, and in an anisotropic material ρ is a tensor, usually given as a matrix function. A linear vector operator is a second-rank tensor.

THREE-ARRAY. See ARRAY.

THRESHOLD. The physical property limit of a phenomenon, particularly the lower limit. For example, the IP saturation threshold would refer to the current density value above which the IP phenomenon becomes nonlinear.

TIME CONSTANT. The time required in an exponentially varying circuit for the voltage to change to $1/e$ (37%) of its original value. Also called DECAY CONSTANT or RELAXATION TIME.

TIME DOMAIN. In circuit theory and measurement, the domain in which time is the variable parameter. In the field of induced polarization also referred to as the PULSE or DC PULSE METHOD.

TORTUOSITY. The inverse ratio of the length of a rock specimen to the length of the path of electrolyte within it.

TRANSFER FUNCTION. A description of system characteristics in the frequency domain. See TRANSFER IMPEDANCE.

TRANSFER IMPEDANCE. A term drawn from circuit theory which can also apply to the complex resistivity of a three-dimensional material, considering resistivity to be a function of frequency. The complex ratio of a potential difference at one pair of terminals or electrodes to the current at the other pair. See TRANSFER FUNCTION.

TRANSFORM. To convert information from one form or domain to another, as with Laplace or Fourier transforms. In this way, time-domain data can be transformed to frequency data and vice versa.

TRANSFORMER. An electrical device for transforming (coupling) alternating current to one or more other circuits at a different current and voltage but at the same frequency.

TRANSIENT. A nonrepetitive voltage or current or pulse change usually of short duration.

TRANSIENT IP METHOD. See PULSE IP METHOD.

TRANSISTOR. An electrical device with three or more terminals using a semiconductor for controlling the flow of current between two terminals by means of current flow between one of these terminals and a third terminal. It performs functions similar to those of a vacuum tube without requiring current to heat a cathode. Ideally, it is a current-controlled current source that operates in only one direction.

TRANSMITTER. In IP surveying, a current waveform generator system used in field surveying. Also called SENDER.

TROUBLESHOOT. To look for the cause of a malfunction.

TRUE RESISTIVITY (or TRUE IP EFFECT). The resistivity or IP effect of a locally homogeneous medium. Compare APPARENT RESISTIVITY.

TUNED VOLTMETER. A voltmeter containing a band-pass filter.

TWO-ARRAY. See ARRAY, POLE—POLE.

TWO-DIMENSIONAL PLOT. A contour plot of depth sounding data (apparent resistivity, metal factor, etc.) as a function of position along a line and electrode separation. See PLOTTED PSEUDOSECTION.

TYPE CURVES. Precomputed IP and resistivity response curves, plotted against electrode interval, used for interpreting field data where horizontal or vertical layering is indicated. Data are usually compared by using transparent overlays. Abscissa and ordinate are usually normalized to be dimensionless. Type curve derivations usually employ the method of images. Also called MASTER CURVES.

U

UNIJUNCTION TRANSISTOR. A transistor made of n-type semiconductive material with a p-type alloy region on one side. Connections are made to base contacts at either end of the n-type material and also to the p region. Primarily used in timing circuits.

V

VACUUM-TUBE VOLTMETER (VTVM). A voltage measuring instrument using electronic circuits to obtain high impedance across the measurement probe so that very little current is drawn. The vacuum tube itself is now largely superseded by solid-state circuitry.

VERTICAL ELECTRIC SOUNDING (VES). See SOUNDING.

VERTICAL PROFILE. See SOUNDING.

VLF. Very low frequency. Electromagnetic radiation between 3 and 30 kHz.

VOLTAGE DIVIDER. See ATTENUATOR.

VOLTMETER. An electrical instrument used to measure the potential differences between points in a circuit. A voltmeter can respond to average, root-mean-square, or peak voltage values of a sinusoidal wave.

W

WARBURG IMPEDANCE. In the electrochemical theory of IP phenomena, the ion diffusion impedance involved in current transfer at an electrode by the faradaic path. It is an indication of the rate of the faradaic process. In most electrical models, the magnitude varies inversely with the square root of the electrical frequency.

WARBURG REGION. On a generalized resistivity spectrum, the steep part of the spectral curve near the inflection.

WATT. A unit of electrical power measured as the product of volt-amperes. One watt is 1/746 horsepower or one horsepower is 746 watts.

WAVE IMPEDANCE. In electromagnetic wave propagation theory, the ratio of electric to magnetic field intensity.

WAVE NUMBER (k). See PROPAGATION CONSTANT.

WIND NOISE. An induced noise voltage in a suspended receiver wire of an IP measuring system caused by wire oscillations in the earth's magnetic field.

Z

ZENER DIODE. A silicon diode in which the breakdown voltage in the reverse direction (Zener voltage) is used for voltage stabilization or voltage reference.

ZERO FREQUENCY. Refers to ac phenomena as extrapolated to the lowest, that is, zero, frequency.

ZETA POTENTIAL. The potential difference across the diffuse layer in an electrolyte.

REFERENCES

Adler, P., 1958. Apparent resistivity cross sections, model results, dipole—dipole coupling. *Mass. Inst. Technol. Rep.*, p.44.

Aiken, C.L.V. and Hastings, D., 1973. Numerical modeling of induced polarization. *11th Int. Symp. on Computer Applications in the Minerals Industry, College of Mines, Univ. of Arizona*, p.H96—H111.

Anderson, L.A. and Keller, G.V., 1964. A study in induced polarization. *Geophysics*, 29: 848—864.

Bacon, L.O., 1965. Induced polarization logging in the search for native copper. *Geophysics*, 30: 246.

Bertin, J., 1968. Some aspects of induced polarization (time domain). *Geophys. Prosp.*, 16: 401—426.

Bertin, J. and Loeb, J., 1974. Traitement 'à la main' et sur ordinateur des transistoires en polarization provoquée. *Geophys. Prosp.*, 22: 93—106.

Bhattacharyya, P.K. and Patra, H.P., 1968. *Direct Current Geolectric Sounding — Principles and Interpretations.* Elsevier, Amsterdam.

Bleil, D.F., 1953. Induced polarization: a method of geophysical prospecting. *Geophysics*, 18(3): 636—662.

Bodmer, R., Ward, S.H. and Morrison, H.F., 1968. On induced electrical polarization and ground water. *Geophysics*, 33: 805—821.

Brant, A.A., 1959. Historical summary of overvoltage developments by Newmont Exploration Limited, 1946—1955. In: J.R. Wait (editor), *Overvoltage Research and Geophysical Applications.* Pergamon Press, London, p.1—3.

Brant, A.A., 1964. Where are we now? *Soc. Min. Eng. Trans.*, 229: 175—183.

Brant, A.A. and Gilbert, E.A., 1952. *Geophysical Exploration, U.S. Patent No. 2611004.* U.S. Patent Office, Washington, D.C.

Coggon, J.H., 1971. Electromagnetic and electrical modeling by the finite element method. *Geophysics*, 36: 137—155.

Cole, K.S. and Cole, R.H., 1941. Dispersion and absorption in dielectrics. I. Alternating current fields. *J. Chem. Phys.*, 9: 341.

Collett, L.S., 1959. Laboratory investigation of overvoltage. In: J.R. Wait (editor), *Overvoltage Research and Geophysical Applications.* Pergamon Press, London, p.50—70.

Dakhnov, V.N., 1959. The application of geophysical methods; electrical well logging. Moscow Petroleum Institute. (English translation by Keller, G.V., 1962. *Colo. School Mines Q.*, 57.)

Delahay, P., 1954. *New Instrumental Methods in Electrochemistry.* Interscience, New York.

Dolan, W.M., 1973. Down-the-hole surveys — geophysical. In: I.A. Given (editor), *Mining Engineering Handbook*, Society of Mining Engineers of AIME, New York, p.5.65—5.70.

Dolan, W.M. and McLaughlin, G.H., 1967. Consideration concerning measurement standards and design of pulsed IP equipment. *Symp. on Induced Electrical Polarization, Eng. Geosc., Dep. Mineral Technol., Univ. of California, Berkeley, 1967, Proc.*, p.2—31.

Elliot, C.L. and Guilbert, J.M., 1974. Induced polarization response attributed to magnetite and certain Fe-Ti oxide minerals. *44th Ann. Int. Meet., Soc. Explor. Geophys., Dallas, Texas, 1974, Abstr., Biograph.*, p.42 (abstract).

Fraser, D.C., Keevil, N.B. and Ward, S.H., 1964. Conductivity spectra of rocks from the Craigmont ore environment. *Geophysics*, 29: 832—847.

Geoscience, Inc., 1968. *Induced Polarization Theoretical Electromagnetic Coupling Curves.* Cambridge, Mass.

Grahame, D.C., 1947. The electrical double layer and the theory of electrocapillarity. *Chem. Rev.*, 41: 3—27.

Grahame, D.C., 1952. Mathematical theory of the faradaic admittance. *Electrochem. Soc. J.*, 99: 370C—384C.

Grant, F.S. and West, C.F., 1965. Interpretation Theory in Applied Geophysics. McGraw-Hill, New York.

Griffiths, D.H. and King, R.F., 1965. *Applied Geophysics for Engineers and Geologists.* Pergamon Press, London.

Grim, R.E., 1968. *Clay Mineralogy.* McGraw-Hill, New York.

Hallof, P.C., 1964. A comparison of the various parameters employed in the variable frequency induced polarization method. *Geophysics*, 29: 425—434.

Hallof, P.C., 1967. An appraisal of the variable frequency IP method after twelve years of application. *Symp. on Induced Electrical Polarization, Eng. Geosc., Dep. Mineral Technol. Univ. of California, Berkeley, 1967, Proc.*, p.51—92.

Hallof, P.C., 1973. The IP phase measurement and inductive coupling. *43rd Ann. Int. Meet., Soc. Explor. Geophys.* and *5th Meet., Asoc. Mexicana Géofis. Explor., Mexico City, 1973, Abstr., Authors' Biograph.*, p.53—54 (abstract).

Hallof, P.H., 1974. Removing inductive coupling effects from IP phase measurements. *44th Ann. Int. Meet., Soc. Explor. Geophys., Dallas, Texas, 1974, Abstr., Biograph.*, p.53—54 (abstract).

Handel, S., 1966. *A Dictionary of Electronics.* Penguin, Baltimore, Maryland, 369 p.

Hauck, A.T. III, 1970. A reconnaissance downhole induced polarization and resistivity survey method. *40th Ann. Int. Meet., Soc. Explor. Geophys., New Orleans, Louisiana, 1970, Geophysics*, 35: 1165 (abstract).

Heinrichs, W.E., Jr., 1967. Discussion of paper presented by G.D. Van Voorhis and R.E. MacDougall, An I.P. case study of a responsive black limestone. *Symp. on Induced Electrical Polarization, Eng. Geosci., Dep. Mineral Technol. Univ. of California, Berkeley, 1967.*

Henkel, J.H. and Van Nostrand, R.G., 1957. Experiments in induced polarization. *Min. Eng.*, 9: 355—359.

Hohmann, G.W., 1972. *Electromagnetic Scattering by Two-Dimensional Inhomogeneities in the Earth.* Dissertation, Univ. of California, Berkeley.

Hohmann, G.W., Kintzinger, P.R., Van Voorhis, G.D. and Ward, S.H., 1970. Evaluation of the measurement of induced electrical polarization with an inductive system. *Geophysics*, 35: 901—915.

Jacobs, J.A., 1970. *Geomagnetic Micropulsations.* Springer, Berlin.

Kaku, H., 1966. On the coupling effect in the induced polarization method. *Butsuri Tanko*, 19 (405): 168—175.

Katsube, T.J. and Collett, L.S., 1973. Electrical characteristic differentiation of sulfide minerals by laboratory techniques. *43rd Ann. Int. Meet., Soc. Explor. Geophys.* and *5th Meet., Asoc. Mexicana Geofis. Explor., Mexico City, 1973, Abstr., Authors Biograph.*, p.54 (abstract).

Keller, G.V., 1959. Analysis of some electrical transient measurements on igneous sedimentary and metamorphic rocks. In: J.R. Wait (editor), *Overvoltage Research and Geophysical Applications*, Pergamon Press, London.

Keller, G.V., 1966. Electrical properties of rocks and minerals. In: S.P. Clark Jr. (editor

Handbook of Physical Constants. Geol. Soc. Am. Mem., 97: 533—577 (revised ed.).

Keller, G.V. and Frischknecht, F.C., 1966. *Electrical Methods in Geophysical Prospecing.* Pergamon Press, London.

Kenney, J.F., Knaflich, H.B. and Liemohn, H.B., 1968. Magnetospheric parameters determined from structured micropulsations. *J. Geophys. Res.*, 73: 6737—6749.

Koefoed, O., 1967. Units in geophysical prospecting. *Geophys. Prosp.*, 15: 1—6.

Koefoed, O., 1968. The application of the kernel fúnction in interpreting geoelectrical resistivity measurements. *Geopublication Associates, Geoexploration Monogr, Ser.* 1, No. 2, Borntraeger, Berlin.

Komarov, V.A., 1969. The importance of induced polarization method for the exploration of ore deposits. In: *Mining and Groundwater Geophysics/1967. Can. Geol. Surv. Geol. Rep.*, 26: 138—147.

Kunetz, G., 1966. Principles of direct current resistivity prospecting. *Geopublication Associates Geoexploration Mon., Ser.* 1, No. 1.

Kunori, S., Yokoyama, H. and Horitsu, T., 1967. The measurement of telluric current activity in the frequency range of 0.01 to 10 cp's. *Butsuri Tanko*, 20: 263—271.

La Compagnie Générale de Géophysique, 1955. *Abaques de sondage electrique.* VIII, Supp. 3. European Association of Exploration Geophysicists, The Hague.

Lazreg, H., 1972. *Master Curves for the Wenner Array.* Inland Waters Directorate, Water Resources, Ottawa.

Madden, T. and Cantwell, R., 1967. Induced polarization, a review. *Min. Geophys.*, 2: 373—400.

Madden, T.R. and Marshall, D.J., 1958. A laboratory investigation of induced polarization. *U.S. A.E.C. Rep.*, RME-3156.

Madden, T.R. and Marshall, D.J., 1959a. Electrode and membrane polarization. *U.S. A.E.C. Rep.*, RME-3157.

Madden, T.R. and Marshall, D.J., 1959b. Induced polarization, a study of its causes. *Geophysics*, 24: 790—816.

Madden, T.R. and Marshall, D.J., 1959c. Induced polarization. *U.S. A.E.C. Rep.*, RME-3160.

Madden, T.R. and Thompson, W., 1965. Low-frequency electromagnetic oscillations of the earth ionosphere cavity. *Rev. Geophys.*, 3: 211—254.

Madden, T.R., Fahlquist, D.A. and Neves, A.S., 1957. Background effects in the induced polarization method of geophysical exploration. *U.S. A.E.C. Rep.*, RME-3150.

Maillet, R., 1947. The fundamental equations of electrical prospecting. *Geophysics*, 12: 529—556.

Maillot, E.E. and Sumner, J.S., 1966. Electrical properties of porphyry deposits at Ajo, Morenci, and Bisbee, Arizona. *Min. Geophys.*, 1: 273—287.

Mathisrud, G.C. and Sumner, J.S., 1967. Underground induced polarization surveying at the Homestake mine. *Min. Congr. J.*, 5(3): 66—69.

Mayper, V., Jr., 1959. The normal effect. In: J.R. Wait (editor), *Overvoltage Research and Geophysical Applications.* Pergamon Press, London, p.125—158.

McEuen, R.B., Berg, J.W. and Cook, K.L., 1959. Electrical properties of synthetic metalliferous ore. *Geophysics*, 24: 510—530.

McMurry, H.V. and Hoagland, A.D., 1956. Three-dimensional applied potential studies at Austinville, Virginia. *Geol. Soc. Am. Bull.*, 67: 683—696.

Millett, F.B., Jr., 1967. Electromagnetic coupling of colinear dipoles on a uniform halfspace. *Min. Geophys.*, 2: 401—419.

Molinski, A.E., 1970. Effects of electricity on the human body. *Min. Cong. J.*, 56(7): 64—68.

Morrison, B.C., 1970. An electrical method adapted for use in a single hole to give direction to a better conductor. *40th Ann. Int. Meet., Soc. Explor. Geophys., New Orleans, Louisiana, 1970; Geophysics*, 35: 1164 (abstract).

Nabighian, M.N. and Elliot, C.L., 1974. Unusual induced polarization effects from a horizontally three-layered earth. *44th Ann. Int. Meet., Soc. Explor. Geophys., Dallas, Texas, 1974, Abstr., Biograph.*, p.52—53 (abstract).

Orellana, E. and Mooney, H.M., 1966. *Master Tables and Curves for Vertical Electrical Sounding over Layered Structures.* Interciencía, Madrid.

Orellano, E. and Mooney, H.M., 1972. *Two and Three Layer Master Curves and Auxiliary Point Diagrams for Vertical Electric Sounding Using Wenner Arrangement.* Interciencía, Madrid.

Parasnis, D.S., 1973. *Mining Geophysics.* Elsevier, Amsterdam, 2nd ed.

Peterson, R.C. and Sumner, J.S., 1973. Computer modeling of data from a downhole IP survey. *11th Int. Symp. on Computer Applications in the Minerals Industry, College of Mines, Univ. of Arizona, Proc.*, p.H.1—H.19.

Roman, I., 1931. How to compute tables for determining electrical resistivity of underlying beds and their application to geophysical problems. *U.S. Bur. Mines Tech. Pap.* 502.

Ruddock, K.A., 1959. Field equipment for prospecting by the overvoltage method. In: J.R. Wait (editor), *Overvoltage Research and Geophysical Applications.* Pergamon Press, London, p.112—114.

Saito, T., 1969. Geomagnetic pulsations. *Space Sci. Rev.*, 10: 319—412.

Salt, D.J., 1966. Tests of drill hole methods of geophysical prospecting on the property of Lake Dufault Mines Limited, Dufresnoy Township, Quebec. In: *Mining Geophysics. I, Case Histories.* Society of Exploration Geophysicists, Tulsa, Oklahoma, p.206—226.

Sauck, W.A., 1969. *A Laboratory Study of Induced Electrical Polarization in Selected Anomalous Rock Types.* Unpubl. M.S. thesis, Univ. of Arizona.

Schillinger, A.W., 1964. Calumet successfully uses IP probe underground to boost ore discoveries. *Min. Eng.*, 16(11): 83—88.

Schlumberger, C., 1920. *Étude sur la prospection électrique du sous-sol.* Gauthier-Villars, Paris.

Schumann, W.O., 1952. Über die strahlungslosen Eigenschwingungen einer leitenden Kugel, die von einer Luftschiehl und einer Ionosphärenhülle umgeben ist. *Z. Naturforsch.*, 7A: 149—154.

Scott, J.H., Carroll, R.D. and Cunningham, D.R., 1967. Dielectric constant and electrical conductivity measurements of moist rock: a new laboratory method. *J. Geophys. Res.*, 72: 5101—5115.

Scott, W.J. and West, C.F., 1969. Induced polarization of synthetic high resistivity rocks containing disseminated sulphides. *Geophysics*, 34: 87—100.

Seigel, H.O., 1948. *Theoretical Treatment of Selected Topics in Electromagnetic Prospecting.* National Research Council of Canada, p.34—53 (unpublished).

Seigel, H.O., 1949. *Theoretical and Experimental Investigation into the Application of the Phenomenon of Overvoltage to Geophysical Prospecting.* Ph.D. Thesis, Univ. of Toronto.

Seigel, H.O., 1959a. A theory for induced polarization effects (for step-function excitation). In: J.R. Wait (editor), *Overvoltage Research and Geophysical Applications.* Pergamon Press, London, p.4—21.

Seigel, H.O., 1959b. Mathematical formulation and type curves for induced polarization. *Geophysics*, 24: 547—565.

Seigel, H.O., 1974. The magnetic induced polarization (MIP) method. *Geophysics*, 39: 321—339.

Sheriff, R.E., 1973. *Encyclopedic Dictionary of Exploration Geophysics.* Society of Exploration Geophysicists, Tulsa, Oklahoma.

Slichter, L.B., 1933. The interpretation of the resistivity prospecting method for horizontal structures. *Physics*, 4: 307—322, 407.

Slichter, L.B., 1955. Geophysics applied to prospecting for ores. In: A.M. Bateman (editor), *Fiftieth Anniversary Volume, 1905—1955.* Economic Geology Publishing Co., Urbana, Ill., Part II, p.885—969.

Sumi, F., 1961. The induced polarization method in ore investigation. *Geophys. Prosp.*, 9: 459—477.

Sumi, F., 1965. Prospecting for non-metallic minerals by induced polarization. *Geophys. Prosp.*, 13: 603—616.

Sunde, E.D., 1949. Earth Conduction Effects in Transmission Systems. Van Nostrand, New York, 373 p.

Swift, C.M., 1973. The L/M parameter of time-domain IP measurements — a computational analysis. *Geophysics*, 38: 61—67.

Tagg, G.F., 1964. *Earth Resistances.* Pitman, London.

The Netherlands Rijkswaterstaat, 1969. *Standard Graphs for Resistivity Prospecting.* European Association of Exploration Geophysicists, The Hague.

Troitskaya, V.A. and Gul'elmi, A.V., 1967. Geomagnetic micropulsations and diagnostics of the magnetosphere. *Space Sci. Rev.*, 7 : 689—768.

Vacquier, V., Holmes, C.R., Kintzinger, P.R. and Lavergne, M., 1957. Prospecting for ground water by induced electrical polarization. *Geophysics*, 22: 660-687.

Van Nostrand, R.G. and Cook, K.L., 1966. Interpretation of resistivity data. *U.S. Geol. Surv. Prof. Pap.*, 499.

Van Voorhis, G.D. and MacDougall, R.E., 1967. An IP case study of a responsive black limestone (by title only). *Symp. on Induced Electrical Polarization, Eng. Geosc., Dep. Mineral Technol., Univ. of California. Berkeley, 1967.*

Van Voorhis, G.D., Nelson, P.H. and Drake, T.L., 1973. Complex resistivity spectra of porphyry copper mineralization. *Geophysics*, 38: 49—60.

Vladimirov, N.P. and Kleimenova, N.G., 1962. On the structure of the earth's natural electromagnetic field in the frequency range of 0.5—100 CPS. *Acad. Sci., USSR, Geophys. Ser. Bull.*, 10. English edition translated by American Geophysical Union, p. 852—855.

Von Hippel, A.R., 1954. *Dielectrics and Waves.* Wiley, New York.

Vozoff, K. 1958. Numerical resistivity analysis horizontal layers. *Geophysics*, 23: 536—556.

Wagg, D.M. and Seigel, H.O., 1963. Induced polarization in drill holes. *Can. Min. J.*, April, p.54—59.

Wait, J.R., 1958. Discussions on a theoretical study of induced electrical polarization. *Geophysics*, 23: 144—154.

Wait, J.R. (editor), 1959a. *Overvoltage Research and Geophysical Applications.* Pergamon Press, London.

Wait, J.R., 1959b. A phenomenological theory of overvoltage for metallic particles. In: J.R. Wait (editor), *Overvoltage Research and Geophysical Applications.* Pergamon Press, London, p.22—28.

Wait, J.R., 1959c. The variable-frequency method. In: J.R. Wait (editor), *Overvoltage Research and Geophysical Applications.* Pergamon Press, London, p.29—49.

Ward, S.H., 1963. Dynamics of the magnetosphere. *J. Geophys. Res.*, 68: 781—788.

Ward, S.H., 1967. Electromagnetic theory for geophysical applications. In: *Mining Geophysics. II, Theory.* Society of Exploration Geophysicists, Tulsa, Oklahoma, p.10—196.

Ward, S.H. and Fraser, D.C., 1967. Conduction of electricity in rocks. In: *Mining Geophysics. II, Theory.* Society of Exploration Geophysicists, Tulsa, Oklahoma, p.197—223.

Ward, S.H. and Rogers, G.R., 1967. Introduction. In: *Mining Geophysics. II, Theory.* Society of Exploration Geophysicists, Tulsa, Oklahoma, p.3—8.

Ware, H., 1973. Geophysical study of the Yerington, Nevada porphyry copper deposit. *Paper presented at the 1973 Ann. Meet. of AIME, San Francisco, 1973.*
White Electromagnetics, Inc., 1963. *A Handbook on Electrical Filters.* Rockville, Maryland.
Wynn, J.C., 1974. *Electromagnetic Coupling in Induced Polarization.* Dissertation, Univ. of Arizona, Tucson. University Microfilms, Ann Arbor, Michigan.
Zonge, K.L., 1972. *Electrical Parameters of Rocks as Applied to Geophysics.* Dissertation, Univ. of Arizona, Tucson. University Microfilms, Ann Arbor, Michigan.
Zonge, K.L., Sauck, W.A. and Sumner, J.S., 1972. Comparison of time, frequency, and phase measurements in induced polarization. *Geophys. Prosp.*, 20: 626—648.

INDEX